Introduction to Automotive Engineering

Scrivener Publishing
100 Cummings Center, Suite 541J
Beverly, MA 01915-6106

Publishers at Scrivener
Martin Scrivener (martin@scrivenerpublishing.com)
Phillip Carmical (pcarmical@scrivenerpublishing.com)

Introduction to Automotive Engineering

R. Sakthivel, Faisal O. Mahroogi,
S. Narayan, S. Abubakar,
M. U. Kaisan and Youssef Alammari

Scrivener Publishing

WILEY

This edition first published 2019 by John Wiley & Sons, Inc., 111 River Street, Hoboken, NJ 07030, USA and Scrivener Publishing LLC, 100 Cummings Center, Suite 541J, Beverly, MA 01915, USA

For more information about Scrivener publications please visit www.scrivenerpublishing.com.

Wiley Global Headquarters
111 River Street, Hoboken, NJ 07030, USA

For details of our global editorial offices, customer services, and more information about Wiley products visit us at www.wiley.com.

Library of Congress Cataloging-in-Publication Data

ISBN 978-1-119-47980-2

Cover image: Traimak Ivan | Bubushonok | Dreamstime.com
Cover design by Kris Hackerott

Set in size of 15 pt and Minion Pro by Exeter Premedia Services Private Ltd., Chennai, India

10 9 8 7 6 5 4 3 2 1

Contents

1

Introduction

1.1 Classification of Motor Vehicle

Motor Vehicles are used for transporting goods or passengers or fulfilling specific functions on land. Motor vehicles can move on the ground, as compared to aircraft and marine craft that operate in air or water.

1.1.1 Based on Type of Roads

1. Guided and Non-guided vehicles
   Guided Motor vehicles move along a fixed guide way; that includes railway vehicles. Non-guided motor vehicles move in any direction. The non-guided motor vehicles are the subject of this book.

2. Classification according to Running gear – Single Track motor vehicles and Multi-Track motor vehicles
   Single Track motor vehicles are motor vehicles with two wheels with or without a sidecar. Examples: Motorcycle – Any two-wheeled vehicle with or without a sidecar.

Moped - Motor cycle with pedals and a petrol engine of low power.

Single-Track Motor Vehicles

Scooter –This is a light, small wheels automotive.

Multi-Track Motor Vehicles -Motor vehicles with three or more wheels.

- Salooncar (sedan) –This is a type of motor car in which the space for driver and passengers is cut off from other areas.
- Hatchback car having a large sloping back.

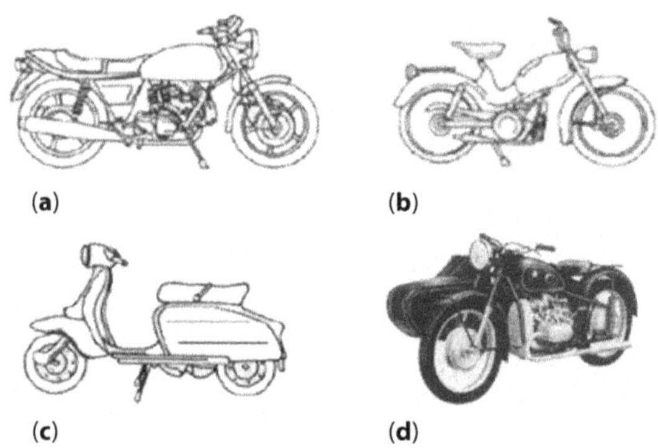

Figure 1.1 Single track motor vehicles.

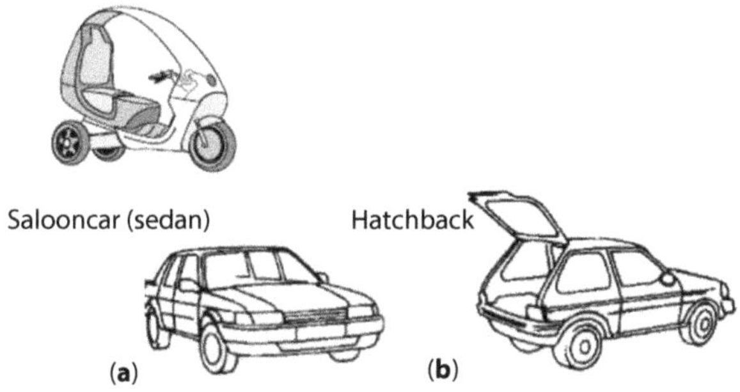

Figure 1.2 Multi-track motors.

- Pick-up (also pickup) – it is a small truck commonly used by operators.
- Van - covered vehicle, with no side windows, for transporting goods or people.
- Truck (UK Lorry) - large strong motor vehicle for transporting goods, soldiers, etc., by road.
- Tractor fitted with an endless belt passing round the wheels of a tractor enabling it to travel over rough ground.
- Car - Motor vehicle intended for carrying a maximum of 9 passengers with luggage.

1.1.2 Buses

Buses are used to transport people. The capacity of buses is more than 9 passengers and luggage. Buses fall into one of the following categories, depending upon the intended use.

1. Microbus: The capacity of microbuses is approximately 25 passengers.

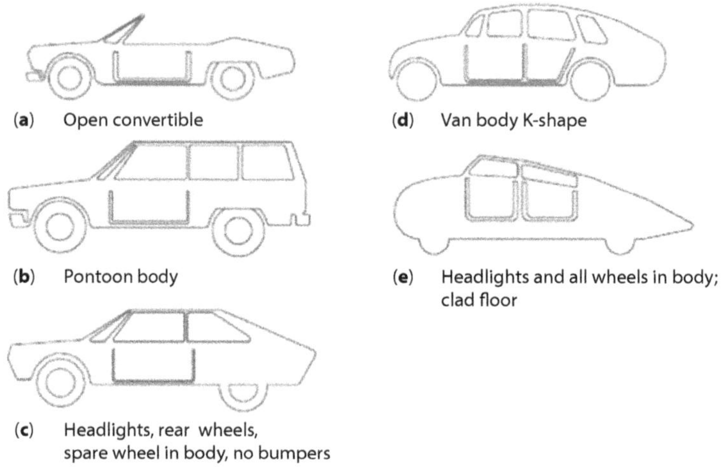

Figure 1.3 Multi-track cars.

2. City buses (Urban buses): are designed and equipped for driving regularly scheduled in-city and suburban routes. Due to short intervals between stops in local traffic, facilitation of rapid passenger turnover by means of low steps, wide doors and as low a vehicle floor as possible, is particularly important.
3. Tour buses (Long-distance coach): are designed to provide a comfortable ride over long distances. Tour buses have large luggage compartments from front to rear below the floor.

Transport Vehicles: Transport vehicles are further classified as follows:

According to the type of body:

1. General-purpose cargo trucks: with an open drop-sided body used for carrying bulk and packaged goods.

2. Special-purpose trucks: with bodies adapted for a certain kind of work, e.g., dump trucks with tiltable bodies used to carry viscous and bulk materials, or with bodies specially adapted for transporting peat, cement, mixed fodder, grain, cotton, cattle, gasoline, milk, etc.

According to the load capacity:

1. Extra-light-duty trucks: up to 0.75 Ton, built on the chassis of passenger cars and used for deliveries of light loads in the communication and communal services and in trade.
2. Light-duty vehicles: from 0.75 to 2.5 Ton, that work in trade, at industrial enterprises and in agriculture, hauling light loads. They also serve as cargo taxis.

According to the load capacity:

1. Medium-duty trucks: from 2.5 to 5.0 Ton, mostly carrying loads for organizations and enterprises with moderate cargo traffic.
2. Heavy-duty trucks: from 5.0 t to 10.0 Ton, used on hard surface roads for carrying building materials, fuel, and products manufactured by large industrial enterprises.
3. Extra-heavy-duty vehicles: above 10.0 Ton, for work in ore and coal mines and at big construction projects with a large amount of steady cargo traffic. As a rule, these vehicles are diverted from public highways.

1.1.3 Classifications by ISO

ISO3833 classifies ground vehicles in 7 groups:

1. Motorbikes
2. Cars for travelers
3. Pick-up Buses
4. Pickup Trucks
5. Agricultural machinery
6. Trailer cars
7. Semi-trailers

Classifications according to their drives as:

1. A 2 wheel Drive
2. Rear-wheel Front engine drive

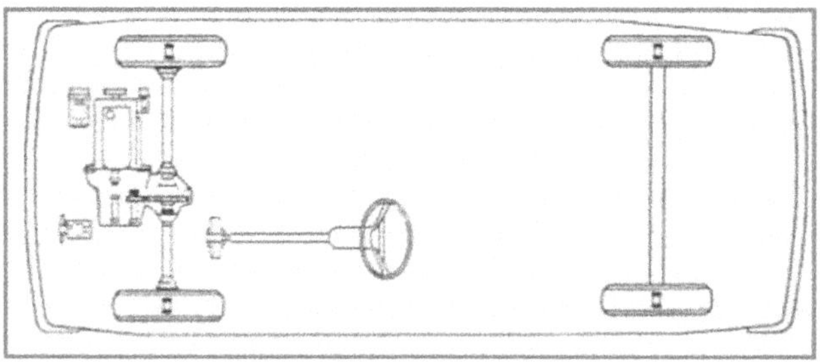

Figure 1.4 Cross-wheel drive.

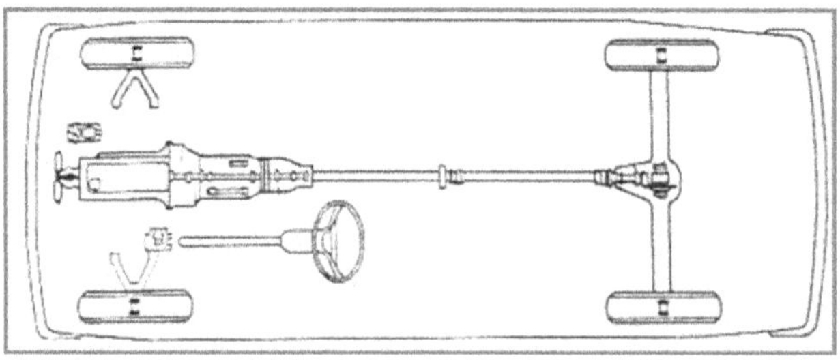

Figure 1.5 Front engine front-wheel drive.

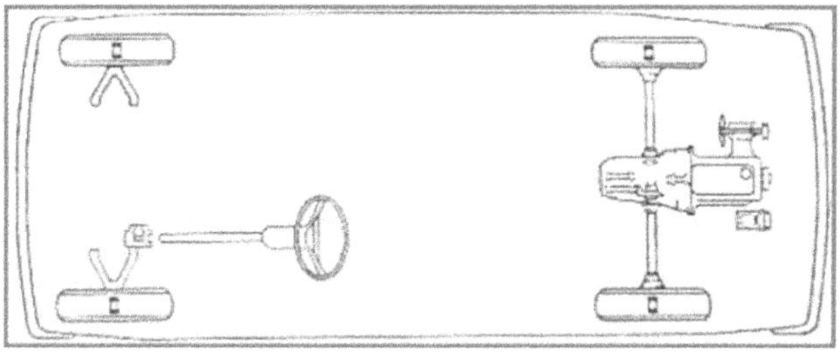

Figure 1.6 Front-wheel rear engine drive.

3. Front-wheel Front engine drive
4. cross mounted
5. Engine behind transmission
6. Engine in front of transmission
7. Rear-wheel Rear engine drive
8. 4-wheel drive

1.2 Functions of Subunits

Mechanical energy is converted into Motion. Any Automobile consists of Power-train and Non-Power-train components. Power-Train components parts involved in the process of generating power to converting it to motion. Non-Power-Train components parts includes Body, Frame and Steering.

Parts of Automobile Systems

1. Power plant
2. Engine
3. Drive train
4. Clutch, gear box, shaft and differential
5. Electrical system
6. Ignition, starting lighting and battery

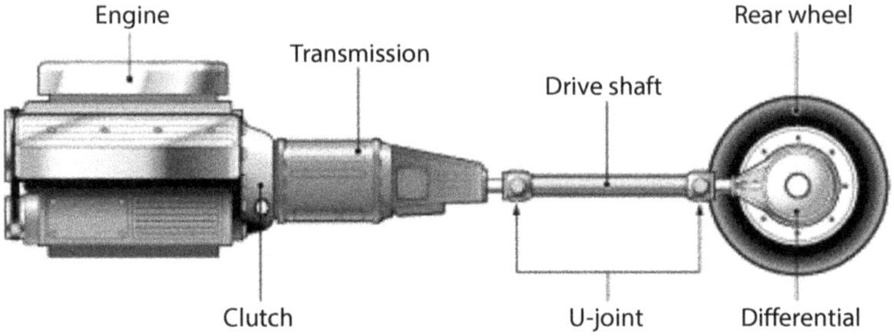

Figure 1.7 Transmission system.

7. Wheels and tires
8. Suspension system
9. Steering system
10. Braking system

Drive train indicates the power flow of motor vehicle from engine to driving wheels.

The main components of the drive train are:

Engine-Clutch-Transmission-Driveshaft-Differential-Wheels.

1.3 Characteristics of Ground Vehicle

The main aim of studying the mechanics of vehicles is to gain the ability to establish rules for the development, design, and selection of vehicles. Performance means the ability of the vehicle to accelerate, overcome obstacles and to stop. Handling is concerned with the driver's response to the vehicle.

Ride is related to the vibrations from ground due to motion.

2

Transmission System

2.1 Introduction

The transmission helps to transmit power from the engine to the wheels. It consists of gears, shafts, and other electrical connections. It is composed of systems shown in Figure 2.1:

2.2 Clutch

This part is used to engage and disengage the engine from drive. It allows the driver to control the power flow between the engine and transmission or transaxle operating by the friction principle.

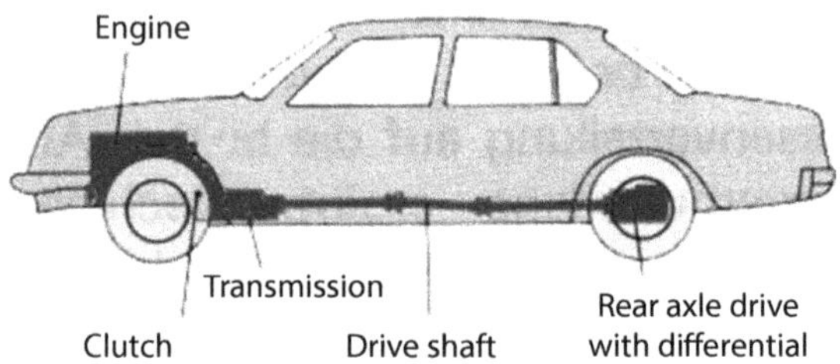

Figure 2.1 Arrangement of components.

Figure 2.2 Engine arrangements.

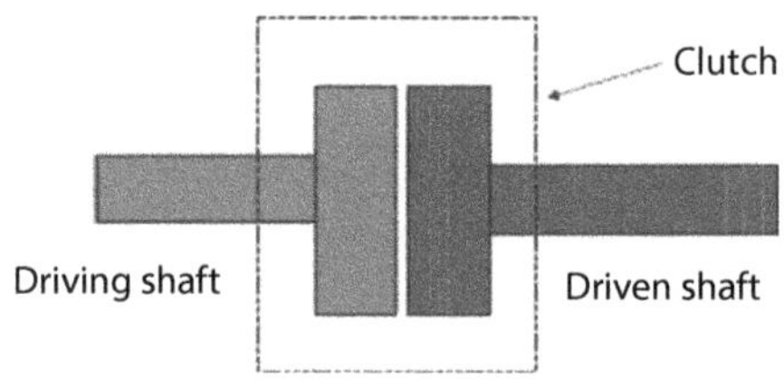

Figure 2.3 Principle of clutch.

Need for Clutch

- Allows gradual engagement of two rotating plates
- Provides positive linkage capable of transmitting maximum engine torque
- Rapidly separates engine from drive train and reengages engine to drive train

- Provides force between pressure plate and flywheel to load clutch disk

Transmission of torque depends on:

- Size
- Engagement time
- Free pedal play
- Rate of Heat loss
- Easy in operation
- Balancing of mass
- Lightness

Power flow from one unit to another can be controlled with a drive disc and a driven disc.

Thus, the clutch can transmit twice as much torque

$$T = 2FR \; T = 2\mu WR$$

For multi-disk clutch T = (n–1)μWR

Where,

W - Spring force (N)

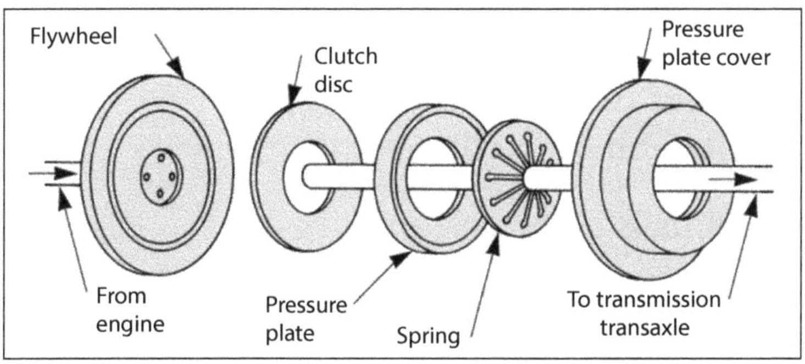

Figure 2.4 Clutch assembly.

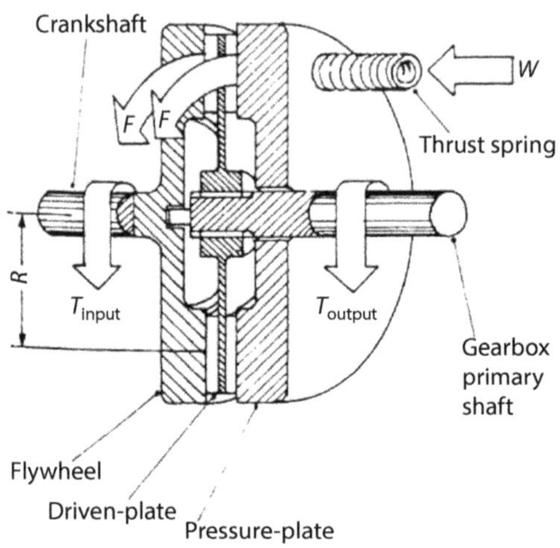

Figure 2.5 Clutch analysis.

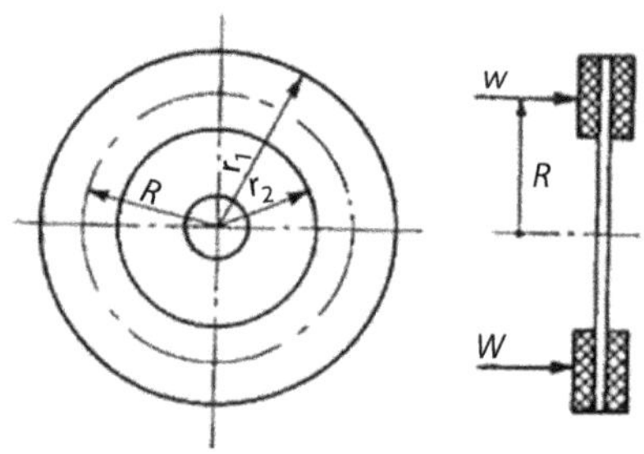

Figure 2.6 Model of clutch.

r_2 - External radius (m)
r_1 - Internal radius of (m)
n - Number of contacting pairs
μ - Coefficient of friction

Mean radius is given as:

$$R = ½\,(r_1 + r_2) = ½\,(r_i + r_o)$$

Tangential force is given by $F = \mu W$

Friction torque is given as $T = F \times R = ½\, \mu W\, (r_1 + r_2)$

Since there are n pairs so:

$$T_F = ½\, \mu n W\, (r_1 + r_2) \qquad \text{in Nm.}$$

If N is speed of the clutch then the power transmitted is given as:

$$P = T_F \times (2\pi N/60) \qquad \text{(Watts)}$$

Example 4.5

A single-plate clutch of 0.15 m effective diameter is lined with material of coefficient of friction 0.35. If the total spring force is 2.5 kN, Calculate:

a. the torque transmitted
b. the power transmitted at 3000 rev/min.

Solution

Total spring force, W = 2.5 kN = 2500 N

Effective radius, R = 0.075 m

Pairs of contact surfaces, n = 2

Coefficient of friction μ = 0.35

Rotational speed, N = 3000 rev/min

a. Torque transmitted, $T_F = \mu W R n$
$= 0.35 \times 2500 \times 0.075 \times 2$ Nm
$= 131.25$ Nm

b. Power transmitted $= T_F \times \frac{2\pi N}{60} W$

$$= 131.25 \times \frac{2\pi \times 3000}{60} W$$

2.3 Synchromesh Gear Box

The arrangement are the same as constant mesh type gear box. Gears on the main shaft mesh with those on lay shaft. The gears on the lay shaft are fixed. This provision avoids the need for double declutching.

Figure 2.7 shows design of a synchromesh gear box. These devices reduce the cost.

Gears *B, C, D, E* in Figure 2.7 are free to move on the main shaft and are in mesh with gears on the lay shaft. Menders F_1 and F_2 are free to move on splines on the main shaft. G_1 and G_2 have internal teeth fit onto the external teeth members $F_1$and F_2 respectively. However, when the force applied in G_1 (G_2) through fork S_1 (S_2) exceeds a certain value, the balls are overcome and member G_1 (G_2) slides over F_1 (F_2).

For direct gear, member G_1 and hence member F_1 (through spring–loaded balls) is slid towards the left till comes M_1 and M_2 rub and friction makes their speed equal. For the second gear the members F_1 and G_1 are slid to the right so that finally the internal teeth on G_1 are engaged with L_1. Then the drive to main shaft will be from *B* via U_1, U_2, *C*, F_1 and splines. For first gear, G_2 and F_2 are moved

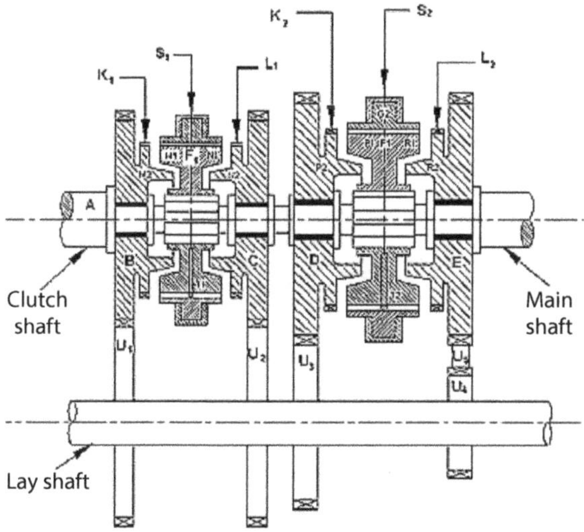

Figure 2.7 Synchromesh gear box.

towards the right. In this case the drive will be from *B* via U_1, U_3, *D*, F_2 and splines to the mainshaft. For reverse, G_2 and F_2 are slid towards the right. In this case the drive will be from *B* via, U_1, U_4, U_5, *D*, F_2 are splines to the main shaft.

2.4 Differential Gear Box

Input torque is sent to the ring gear and then to carrier. The carrier gear is connected to both the side gears by the planet gear. The planet gear revolves driving the side gears.

Torque to the Propeller shaft is supplied by transmission line. A bevel gear takes drive from propeller shaft, and is encased within the housing. This meshes with ring gear, also called crown wheel. If the left side gear encounters resistance, the planet gear spins as well as revolving, allowing the left side gear to slow down.

2.5 Planetary Gear Systems

Figure 2.8 shows the basic layout of the planetary gear system. This consists of four elements as follows,

- Sun Gear – it is located at the center. Input shaft is connected to sun gear.
- Planet Gear – Planet gears are engaged with sun as a pinion in the system and they revolve around the sun gear.
- Ring Gear – Ring gear is the boundary of an envelope defined for the packaging of the sun and the planetary gear
- Arm – Shaft of sun gears and planet gears are connected internally with arms

Figure 2.8 Differential gear box.

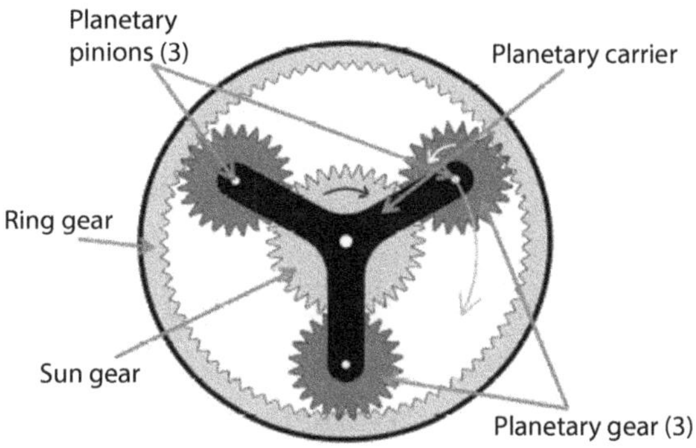

Figure 2.9 Planetary gear system.

2.6 Manual Transmission System

Stages

Fully depressed Clutch:

The clutch is completely disengaged during the full depression stage and no torque is transmitted.

Clutch slips:

The clutch slip allows the engine rotation to adjust to the newly selected gear ratio gradually.

Released Clutch:

Torque will be transmitted to the transmission.

2.7 Automatic Transmission System

This advanced transmission system is composed of three major elements namely, (1) epicyclic gear arrangement, (2) torque converter and (3) fluid coupling.

Stages of Automatic Transmission:

- Park (P): lock the transmission.
- Reverse (R): reverse gear, allowing backward motion.
- Neutral (N): disconnects the transmission from the wheel.
- Low (L): The reduction of speed during the traction of elevated land structures can be achieved by selecting the Low mode.
- Drive (D): selecting drive mode

2.8 Propeller Shaft and Drive Shaft

Drive shafts must therefore be strong enough to bear the stress, whilst avoiding too much additional weight as that would in turn increase their inertia.

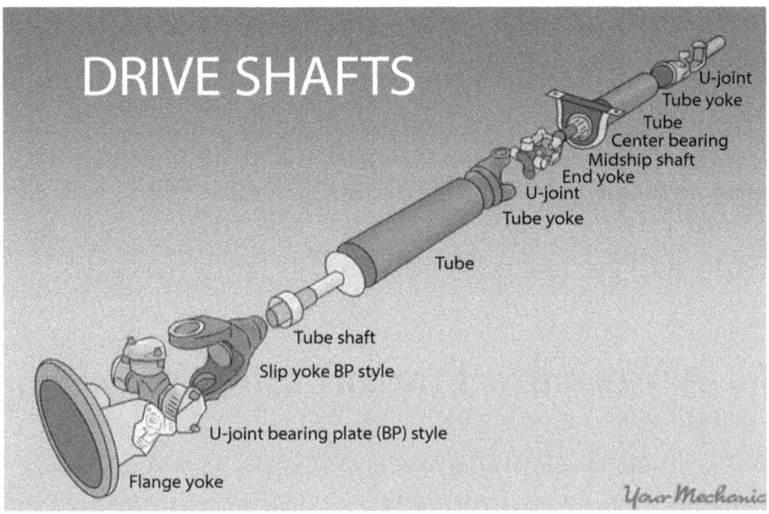

Figure 2.10 Drive shafts.

2.8.1 Role of Propeller Shaft

The torque from the engine is to be transmitted to the rear wheels for propelling vehicle. The drive shaft must provide a smooth, uninterrupted flow of power to the axles.

2.8.2 Functions of the Drive Shaft

a. Torque transmission.
b. Maintain speed over roll over bumps in the road.
c. The length changes of shaft by means of a slip joint

3

Tires

3.1 Introduction [1]

The tires are one of the vital functional elements of the automobile for transmission of various forces (vertical, lateral and longitudinal) between the automobile and the ground. The properties of the tires must be predictable by the driver and be constant throughout the traction. Most of the modern applications use pneumatically inflated tires to provide better cushioning effects and to improve shock absorbing capacity. The designing of tires should cover wider aspects like the vehicle's static loading, dynamic loading and the vibration damping capacity, etc. Modern tires are generally made of components like natural and synthetic rubbers, fabrics along with carbon black and some organic chemical compounds. Before utilizing rubber tires, the wooden wheels were wound by metal wires to prevent damage due to wear and tear. In modern days, the rubber tires were inflated by pneumatic supply whereas solid rubber tires were still used in various applications like carts, lawn movers, casters, etc.

3.2 Construction of Tire

A tire provides a cushion between the vehicle and the road, reducing shocks. This effect is provided by compressed air present in the tire. Radial ply tires are best for use.

3.2.1 Tire Tread Designs

Tires generally fall into one of the following categories:

- Directional
- Non-directional
- Symmetric and Asymmetric.

3.2.2 Cross-Ply Tires [2]

In developed countries, cross-ply tires are no longer in use for commercial applications as original or replacement tires. In the present scenario, cross-ply tires are used only as spare tires for automobiles for temporary purposes, in motor cycles, race cars and agricultural vehicles. The cross-ply tire has a supporting framework which consists of at least 2 layers of rubberized fibers which have a bias angle of 20–40°. Based on the strength requirements, even steel cords can be employed in the construction. At the feet of the tire, the layers are wrapped around the tire core on either sides with the folded end of the plies forming the bead. The bead must transfer braking moment and provide permanent seat to the tire. Protective moldings are designed on the sides to prevent the damage from contact with curbstones.

3.2.3 Radial Ply Tires

Radial ply tires have much more flexible sidewalls due to their construction. They use two or more

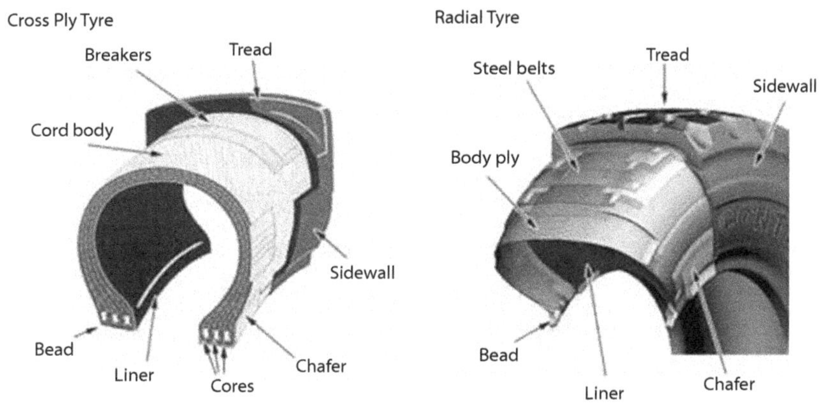

Figure 3.1 Tire construction.

layers of casing plies, with the cord loops running radially from bead to bead.

3.3 Tire Dynamics

At present, two groups of models can be identified, handling models and structural or high frequency models. Structural tire models like RMOD-K or F Tire are very complex. These models are computer time consuming and they need a lot a data. Handling models like the MF-formula or TMeasy rely also on measured and observed force-slip characteristics.

3.3.1 Steady State Tire Forces and Torques

For the calculation of the contact patch geometry the tire is considered as a rigid body. Then, the tire deformation, the orientation of the local contact area, the location of the contact point, and the contact point velocities can be calculated from the momentary state of the wheel rim and the description of the road surface. The rolling resistance torque T_y which is less important for vehicle dynamics may be approximated by a rolling resistance lever. Within handling models the steady state tire forces in longitudinal and lateral direction are approximated by appropriate functions

$$F_x^S = F_x^S\left(s_x, s_y, \ldots\right) \tag{3.1}$$

$$F_y^S = F_y^S\left(s_x, s_y, \ldots \right) \tag{3.2}$$

which mainly depend on the longitudinal and lateral slip s_x and s_y. The steady state torque T_z^S around an axis perpendicular to the local road plane consists of the self aligning torque $T_S^S = c_o F_y^S$ and the bore torque T_B^S

$$T_z^S = c_o F_y^S + T_B^S\left(s_B \ldots \right), \tag{3.3}$$

where $c_o = c_o\ (s_y)$ names the tire caster and s_B denotes a bore slip.

3.3.2 Simple Dynamic Extension

Measurements show that the dynamic reaction of the tire forces and torques to disturbances can be approximated quite well by first order systems. Then, the dynamic tire torque T_z^D are given by first order differential equations

$$\tau_x \dot{F}_x^D + F_x^D = F_x^S \tag{3.4}$$

$$\tau_y \dot{F}_y^D + F_y^D = F_y^S \tag{3.5}$$

$$\tau_\psi \dot{F}_z^D + F_z^D = F_z^S \tag{3.6}$$

which are driven by the steady values F_x^S, F_y^S and T_z^S. The tread particles of a rolling tire move with the transport velocity $r_D|\Omega|$ through the contact patch, where r_D and Ω denote the dynamic rolling radius and the angular velocity of the wheel. Now, time constants τ_i, can be derived from so called relaxation lengths r_i

$$\tau_i = \frac{r_i}{r_D \left|\Omega\right|} \qquad i = x, y, \psi. \tag{3.7}$$

But it turned out that these relaxation lengths are functions of the longitudinal and lateral slip s_x, s_y and the wheel load F_z, Figure 2. Therefore, constant relaxation lengths will approximate the real tire behavior in zero order approximation only. An appropriate model for the dynamic tire performance would be of great advantage because then, the cumbersome task of deriving the relaxation lengths from measurements can be avoided.

References

1. **Tires Wikipedia, Web address:** https://en.wikipedia.org/wiki/Tire Accessed on 18-04-2018
2. **CDX handout activity, Web address:** http://docplayer.net/37315374-Handout-activity-ha494.html Accessed on 18-04-2018.

4

Suspension System

4.1 Introduction

Suspension is the system that connects a vehicle to its wheels and allows relative motion.

4.2 Types of Suspension System

There are three types of height adjustable suspension which are:

- Mechanical
- pneumatic
- hydraulic system

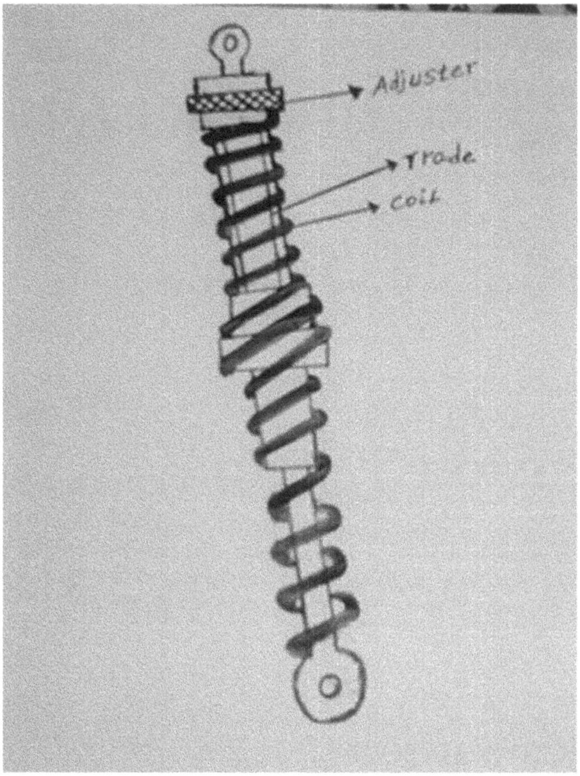

Figure 4.1 Mechanical adjustable concept.

4.2.1 Mechanical System

This system needs manual adjustment to which can be used to vary a vehicle ride height.

4.2.2 Pneumatic System

Air has been used on vehicles at varying frequencies and with mixed success. Air systems can also be used as a means of achieving height adjustment in the vehicle. Air suspension has an added advantage in that the controller can also perform the task of leveling the car if height sensors are placed in each spring. Existing air springs do have the drawback that they cannot simply replace a coil spring in all their behavioral properties. While the properties of air suit heavy vehicle transport they are not as easily put to work underneath a light automobile. Their lack of popularity within smaller vehicles is also due to their elevated cost when compared to a coil spring.

4.2.3 Hydraulic System

The hydraulic system uses hydraulic to vary the height of the vehicle. The system uses pressurize fluid; due to different mechanism it can be by pump or pedal or other mechanism. Hydraulic system has no leaky seals, no variance with temperature or elevation, and it has no noise since it is free from compression.

This condition makes hydraulic the most appropriate when we compare it with other mechanisms.

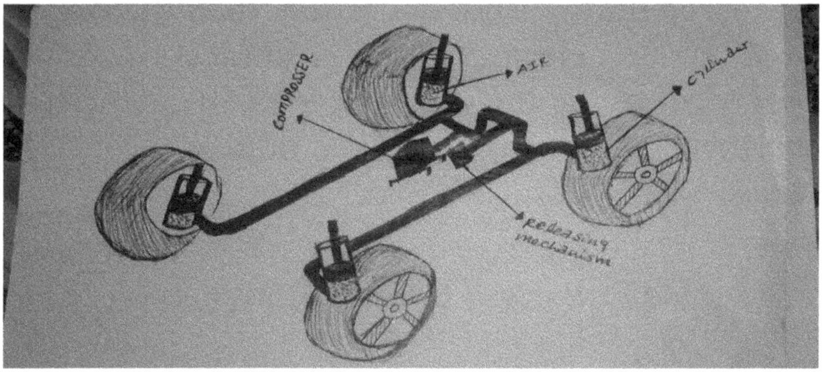

Figure 4.2 Pneumatic adjustable concept.

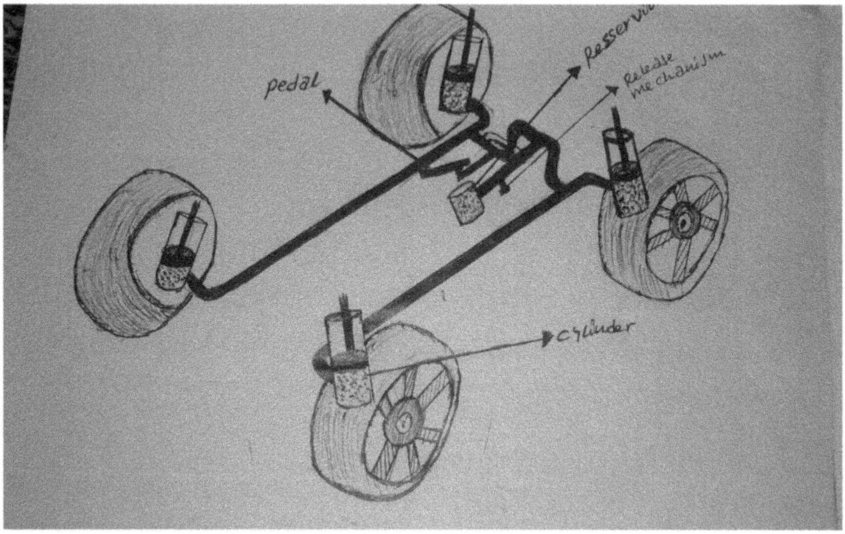

Figure 4.3 Hydraulic adjustable concept.

4.2.3.1 *Working Principle of Hydraulic Suspension System*

When the pedal is pressed repeatedly by the driver, the fluid is sucked from the reservoir tank then pressurized into the main cylinder through the tube. The pressurized fluid pushes the piston, then the vehicle increases ride height from the ground. If the driver needs to decrease the ride height of the vehicle he simply opens the valve then the fluid comeback to the reservoir tank.

4.2.3.2 *Major Parts of Hydraulic Suspension System*

- Pedal
- Main cylinder

Table 4.1 Concept scoring.

	Concept variant			
Selection criteria	**Hydraulic**	**Pneumatic**	**Mechanical**	**Ref**
Noiseless	+	-	0	0
Leaky less	+	-	+	0
Cost	0	-	+	0
Easy	+	+	-	0
Efficiency	+	0	-	0
Plus	4	1	2	0
Minus	0	3	2	0
Same	1	1	1	0
Net	4	-2	0	0
Rank	1	3	2	0
Continue	Yes	No	Yes	

Table 4.2 Concept screening: (based on this table hydraulic system is used).

		Hydraulic		Mechanical	
Selection criteria	**Weight**	**Rating**	**Weighted score**	**Rating**	**Weighted score**
Easy to operate	25%	5	1.25	1	0.25
Efficiency	35%	4	1.4	3	1.05
Time saver	15%	5	0.75	3	0.45
Cost	15%	2	0.3	5	0.75
Noise less	5%	5	0.25	4	0.2
Life time	5%	4	0.2	3	0.15
Total score rank			4.15		2.85
Rank		1		2	
Continue		Yes		No	

- Piston
- Actuator cylinder
- Plunger
- Reservoir cylinder
- Spring
- Ball valve

4.3 Design Analysis of Suspension System

Ground clearance	20 cm
Kerb weight	1270 kg
Gross weight	1900 kg
Overall height	1450 mm
Overall length	4505 mm
Overall width	1755 mm
Tires	195/65/R15

Figure 4.4 3D view of hydraulic suspension vehicle.

Figure 4.5 Model number of the car that we are working on, Honda Civic.

4.3.1 Dimension and Force Analysis of the Pedal

Hydraulic system uses positive displacement input and output device require force analysis. Variation in length and points of supports on this vehicle pedal gives various outputs; thus iteration for acceptable range of length will be important.

Case 1: when the pump piston or plunger is at its lowest position

Where piston is the force that the driver applies on the pedal.

R1- reaction force b/n the link and pedal
R2- reaction force b/n the plunger and the pedal

The pedal socket can be at equilibrium when the force and the momentum should be balanced.

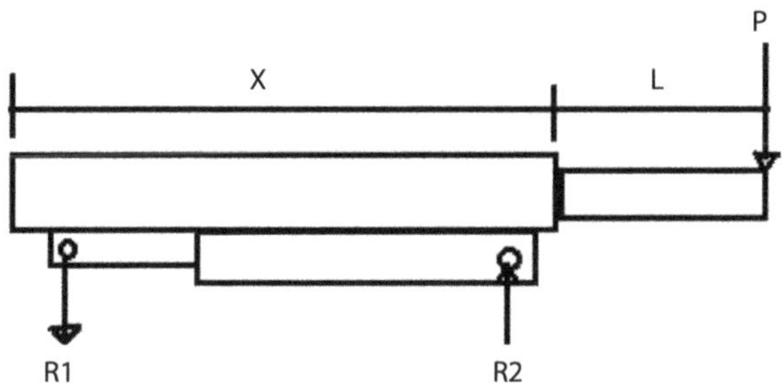

Figure 4.6 Reaction force for pedal.

- $\sum F_y = 0 R_2 = R_1 + P$
- $\sum M_A = 0 \;\; R_2(X) - P(X + L) = 0$

$$R_2(X) = P(X + L)$$

$$R_2 = P\left(\frac{X + L}{X}\right) \quad (4.1)$$

<u>Case 2:</u> when the plunger is at its maximum position

- this happens when the operator up strokes or moves the pedal rise up; this can be shown diagrammatically as follows
- this is the maximum height of the plunger and socket

Where h_A – the distance through which the piston in the pump moves.

θ – The maximum angle through which the pedal moves with horizontals.

Now, $\sin\theta = \frac{h_A}{X}$

$$h_A = X\sin\theta \tag{4.2}$$

Considering Pascal's low of the fluid or oil in the cylinder of the vehicle

- When valve 1 is closed then the oil's route or path is actuating the cylinders
- Valve 2 actuating pipe
- Main cylinder
  - Fr this route or path the pressure is constant all over the path(if valve 1 is closed)
  - Assuming that the typical oil or fluid is incompressible we can apply Pascal's law to the system (the pistons and the path of the oil above)

If F is the load acting on the piston and R_2 be the force acting on the plunger actuator (piston) then

$$F/4 = PAF = (PA)$$

Where p- is the pressure require in the fluid require to hold the piston in equilibrium and

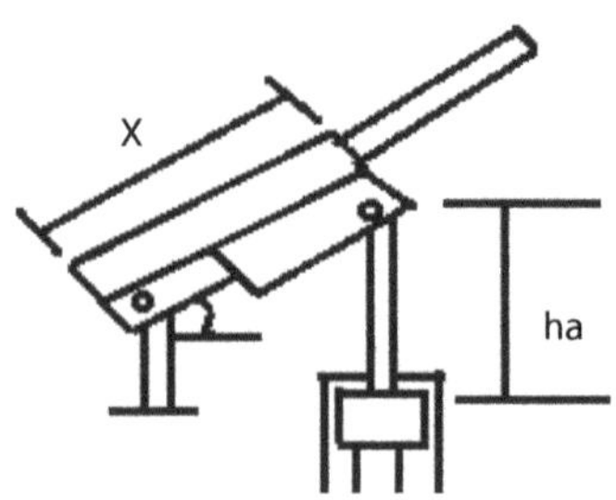

Figure 4.7 Maximum position of plunger.

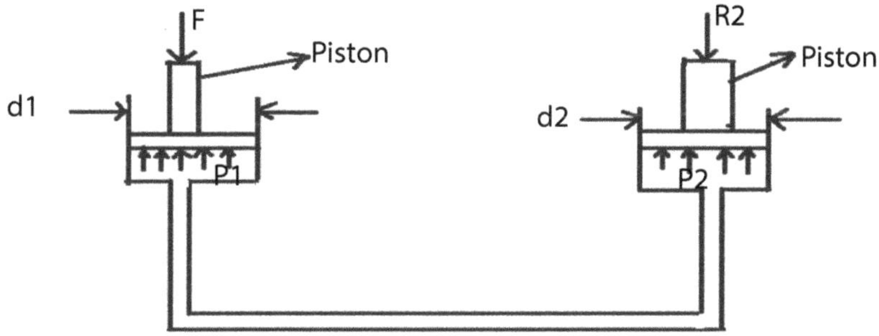

Figure 4.8 Pascal's law.

A- The area of the main cylinder

The force necessary to keep the actuator piston in equilibrium will be

$R_2 = P_2 A_2$ Where A_2 area of the actuator cylinder

From Pascal's law $P_2 = P_1$ thus from the piston

$$P_1 = F/A_1$$

And from the actuating cylinder $P_2 = R_2/A_2$

$$P_1 = P_2 \; F/A_1 = R_2/A_2$$

$$R_2 = \left({}^{A_1}\!\!\diagup\!{}_{A_2} \right) F$$

$$R_2 = \left(\frac{\frac{\pi d_2^{\,2}}{4}}{\frac{\pi d_1^{\,2}}{4}} \right) F$$

$$R_2 = \left(\frac{d_2^2}{d_1^2} \right) F, \text{ where } F = 20KN$$

$$R_2 = 20\left(\frac{d_2}{d_1} \right)^2 \qquad (4.3)$$

- Considering volume relation of the cylinders
- In order to bring the piston to its maximum possible position the actuator cylinder must be moved' times up and down n is in other words called the number of stroke needs to which the maximum lift height volume of cylinder filled to move the piston

= n (volume of cylinder moved by the actuating piston)

$$V_1 = n\,V_2 \qquad n = V_1/V_2$$

Where, V_1 - volume of main cylinder

V_2- volume of the actuating cylinder when the plunger up to its maximum position

$$V_1 = \frac{\pi d_1^2}{4} * \text{height occupied by oil}$$

$$V_1 = \frac{\pi d_1^2}{4}(h_1 - tc - hp)$$

$$V_1 = \frac{\pi d_1^2}{4}(27 - \text{tc} - 2)$$

$$V_2 = \frac{\pi d_2^2}{4} * h_A$$

$$n = V_1 / V_2$$

$$n = \frac{\frac{\pi d_1^2}{4}(27 - tc - 2)}{\frac{\pi d_2^2}{4} * h_A}$$

$$n = \frac{d_1^2 (27 - tc - 2)}{d_2^2 * h_A} \tag{4.4}$$

Now we have got four equations containing the unknown parameter of the geometry.

Now we can summarize the equation as follows,

$$R_2 = \frac{P(X + L)}{X}$$

But this holds true when the pedal is a horizontal position; when the pedal is deflected by an angle θ then the reaction force R_2 is given by

$R_{2\,-}$ Is actuating vertical reaction
P – Is always perpendicular to the pedal axis

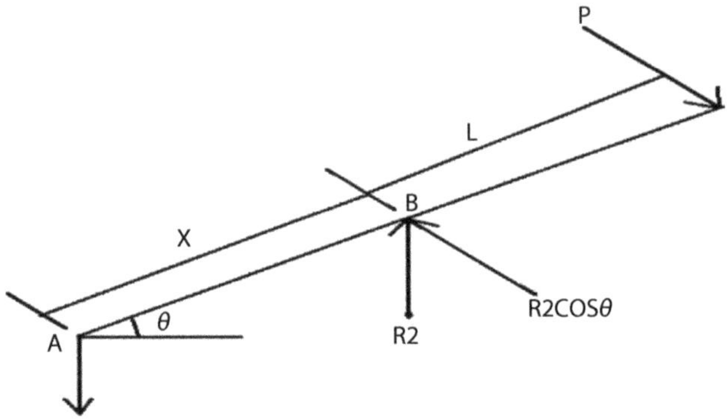

Figure 4.9 Force analysis for pedal.

- Now taking summation of moment due to all force at A will be zero

$$\sum M_A = 0 \quad R_2 cos\theta(x) - p(x + L) = 0$$

$$R_2 cos\theta(x) = p(x + L)$$

$$R_2 = \frac{P(X + L)}{XCOS\theta}$$

When the pedal is at the horizontal position then then the reaction force becomes then $\theta = 0$ then the reaction force becomes

$$R_2 = \frac{P(X + L)}{XCOS\theta} = \frac{P(X + L)}{X}$$

Which is the same with equation (1) derived earlier

- Thus, the reaction force R_2 can more generally be written as

$$R_2 = \frac{P(X+L)}{XCOS\theta} \text{ For any angle } \theta$$

Let us summarize the equations,

$$R_2 = \frac{P(X+L)}{XCOS\theta} \quad (1)$$

$$h_A = xsin\theta \quad (2)$$

$$R_2 = 20\left(\frac{d_2}{d_1}\right)^2 KN \quad (3)$$

$$n = \frac{d_1^2(27 - tc - 2)}{d_2^2 * h_A} \quad (4)$$

D_1=d_P + 2(lower thickness of piston) + 2(clearance b/n the piston head and cylinder)

$$d_1 = d_P + 2(2) + 2(1.5)$$

Where, d_1 - internal diameter of main cylinder

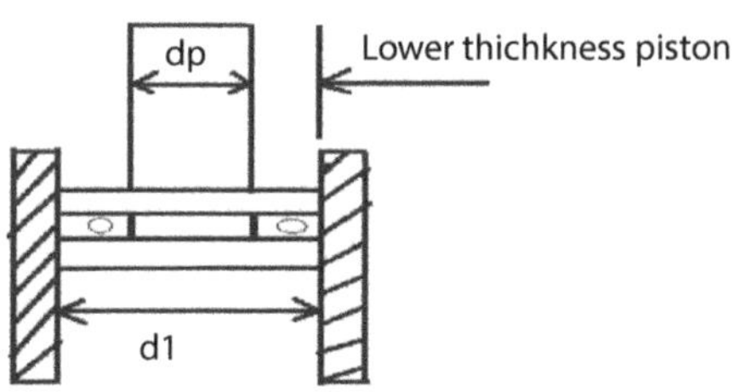

Figure 4.10 Connection of piston and cylinder.

- Let the range of the force the driver can apply on the pedal vary from 200–400N
- Length of the pedal and socket
- Let the length of the pedal L be vary from 400–600mm
- Let the length of the socket b/n the hole of the pin joint X vary from 30–40mm

- Now the driver wants to apply a force of 260N on the pedal the iteration will produce as follow with an objective of finding the possible number of strokes to practically raise the piston up to the indicated maximum lift height

$$R_2 = \frac{260(X+L)}{XCOS\theta}$$

$$d_2 = d_1\sqrt{\frac{R_2}{2000}}$$

$$h_A = xsin\theta$$

$$n = \frac{d_1^2(27-tc-2)}{d_2^2 * h_A}$$

$$n = \frac{d_1^2(27-tc-20mm)}{d_2^2 * h_A}$$

$$n = \frac{d_1^2(27-8-20)}{d_2^2 * h_A}$$

$$n = \frac{d_1^2(242)}{d_2^2 * h_A}$$

As can be seen from Table 4.3, we made five iterations to calculate the unknown over the indicated range of the assumed value, although many designers select a different iterative approach objective. We have selected the optimum and frequently observed number of strokes to be our objective of the iteration process looks the two extremes of the iteration, that is, iteration number 1 and number 2. The first one is too much to be used for 20 KN load and it is mostly stroke number for loads high in magnitude and number 5 is too small to raise the load, thus the 2nd iteration contains the most practicable number of stocks, that is n = 22.2. Then the processing design analysis of each part will depend on the value of parameters obtained from the 3rd iteration.

Generally, the geometry analysis of this system can be summarized,

Height of the main cylinder	27cm
Height that the vehicle rises	20cm
The applied force	260N
Diameter of main cylinder	10cm
Angle that the pedal makes with horizontal	35°
Diameter of actuator cylinder	60mm
Maximum rise of the plunger	15.67mm
Vertical plunger pin reaction force	4.87KN
Number of stroke	$22.2 \approx 23$

Table 4.3 Iteration.

	Known		Assumption					Calculated value			
No iter	H min	H max	P	L	X	d_1	θ	$d_2 = d_1\sqrt{R_2}/20$	$h_A = x\sin\theta$	$R_2 = (X+L)p/X\cos\theta$	$n = d_1^2 * 242/d_2^2 * hA$
1	270	470	260	400	30	100	30	58.6	13.6	4.3	30
2	270	470	260	450	30	100	35	60	15.67	4.87	22.2
3	270	470	260	480	30	100	40	62.5	17.63	5.46	17.07
4	270	470	260	500	30	100	45	70.6	19.48	6.04	13.65
5	270	470	260	550	30	100	50	75.4	21.2	7.1	10.6

Total Height of the piston = head of the piston + height of piston

= 2+20 =22cm

4.3.2 Design of Piston and Piston Head

Piston rod is made of high tensile Stainless steel materials with chromium plating to avoid corrosion and wear. The selection of materials depends on the buckling of the piston rod.

Material selection;

Grade ASTM-A36
Yield strength $\sigma_y = 250$ Mpa
Shear strength $\tau = 145$ Mpa
Modulus of elasticity = 200Gpa

- We have a reason to select this material
  - High compressive strength to support the load
  - Due to its appreciable hardness
  - Give good surface finish
  - Relatively low cost to other grade of steel

Objectives

To check the safety of the piston by comparison of critical load with applied load.

When $F < F_{cr}$ the design is safe

Design procedure

The piston rod may fail in two ways:

- Fail due to compressive stress (crushing)
- Failure due to instability (buckling)

  The cross section of the road critical can be calculated on the following criteria:

- If the length of the rod to least cross-sectional dimension ratio is less than or equal to 11.Then the piston rod is considered as stressed; otherwise it is considered as column.

  $L/d \leq 11$ Short column

  $L/d \geq 11$ Long column

Also, the piston is round shaped d can be substituted b;

$$d = k\sqrt{12} \text{ Where k is slenderness ratio}$$

$$L/k\sqrt{12} \leq 11$$

$$L/k \leq 11\sqrt{12}$$

$$L/k \leq 40$$

- If the length of the slenderness ratio is less than 40 so it is short

$$K = \sqrt{\frac{I}{A}} \quad \text{Where } I = \pi d_p^4 / 4 \times 16$$

$$I = \pi d_p^4 / 64, \quad A = \pi d_p^2 / 4$$

$$K = \sqrt{\frac{\pi d_p^4 / 64}{\pi d_p^2 / 4}} = \sqrt{\frac{\pi d_p^2}{16}}$$

$$K = \frac{d_p}{4}$$

$$L/k \leq 40 \quad L\Big/\frac{d_p}{4} \leq 40$$

$$\text{Thus} \quad \frac{L}{d_p} \leq 10$$

But length of the piston is L=22cm

Assuming the diameter of piston is:

$$d_p = 8\text{cm}$$

$$22/8 \leq 10 \quad 2.75 \leq 10$$

So, the piston is short and will fail due to compressive stress

- Next, we have to check the buckling of the piston
- Crushing of the piston

The maximum load F = 5KN

$\sigma_{all} = \sigma_y / n$ Where n is factor of safety

σ_{all} = 250Mpa/n Where n = 2

σ_{all} = 250 Mpa/2 = 125Mpa

Now considering the crushing strength of the piston

$\sigma_{all} = F_{cr}/A_{cr}$ Where A_{cr} = critical cross sectional area

125 Mpa = $F_{cr}/((\pi dp^2)/4)$ Where dpispiston diameter F_{cr}=critical load

$$F_{cr} = 125\text{Mpa}\ (\pi dp^2)/4$$

$$F_{cr} = 125\text{Mpa}\ (\pi 8^2)/4$$

$$F_{cr} = 6.28\ \text{KN}$$

Now we can conclude from this,

$$F_{cr} > \text{F} > 6.28 > 5\text{KN}$$

Now this implies that 8 cm piston diameter can sustain up to 6.28KN load. This piston can sustain the 5KN load to lift. Now it is possible to take the 8cm piston diameter; now we can design the cross section of the piston to sustain the 5KN maximum load.

$$\sigma_{all} = \frac{F_{\max}}{\pi d_p^2/4} d_p = \sqrt{\left(\frac{4F_{\max}}{\pi \sigma_{all}}\right)}$$

$$d_p = \sqrt{\left(\frac{4 * 5 * 10^3 N}{125 * \pi}\right)}$$

$$= 7.13 \approx d_p = 7cm$$

So, we can take the new diameter of the piston to all other parameters found during the geometry analysis with depend on the geometry of the piston rod.

Design analysis for buckling

The length of the piston rod for buckling design.

$$h_p = 22 \text{ cm}$$

As we have seen before, the piston is considered as a short column. Now Johnson equation

$$F_{cr} = A_{cr} \times \sigma_y \left[1 \frac{\sigma_y}{4c\pi 4c\pi^2 E} \left(\frac{L}{k} \right)^2 \right]$$

Where $A_{cr} = \pi d_p^2 / 4 = \pi 7^2 / 4 = 38.48 \, \text{cm}^2$

$$\sigma_y = 250 \text{ Mpa}$$

$$\text{E} = 200 \text{ Gpa}$$

$$L/K = \text{slenderness ratio}$$

$$K = \sqrt{\frac{\pi d_p^4 / 64}{\pi d_p^2 / 4}} = \sqrt{\frac{\pi d_p^2}{16}}$$

$$K = \frac{d_p}{4}$$

$$\text{K} = 7\text{cm}/4 = 1.75 \text{ cm}$$

$$\text{Thus, } \frac{L}{k} = \frac{22}{1.75} = 12.6$$

C=2 for one end fixed and other pivoted since the piston is kept over the fluid in one end which is move free and the other end is mounted fixed

$$F_{cr} = 38.48\,\text{cm}^2 \times 250\,\text{Mpa}$$
$$\times\left[1 - \frac{250\,Mpa}{4\times 2\times \pi^2 \times 200\times 10^3}(12.6)^2\right]$$
$$= 9620[1 - 0.0025]$$
$$= 9.57\,\text{KN}$$

This is the critical load for buckling so the piston rod will never fail due to buckling for the given F = 5KN load

4.3.3 Piston Head Design

The head of the thickness is taken as a uniform circular flat plate. The pressure of the oil acts up in the plate uniformly. We can select the material that

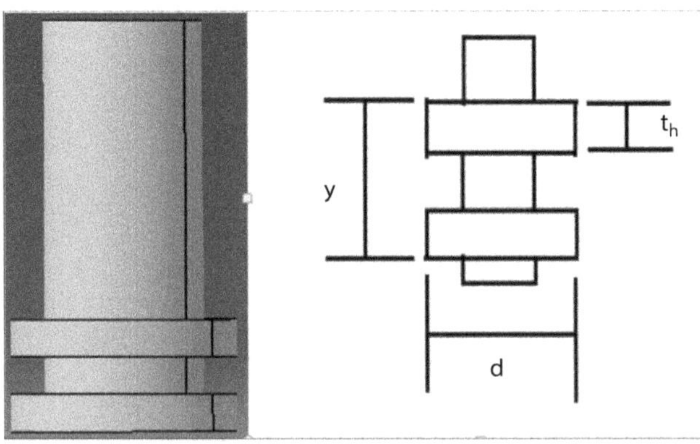

Figure 4.11 Piston head.

we have used in the piston rod using the previous mechanical property.

Design analysis of piston head

We can use the following equation to find the piston head thickness

$$t_h = \sqrt{\left(\frac{3pd_p^2}{16\sigma_{all}}\right)}$$

$t_h = 0.433 d_p / \sqrt{(p/\sigma_{all})}$ Where p = maximum pressure in cylinder

$$t_h = .433 \times 7\text{cm} \times \sqrt{\left(\frac{p}{125Mpa}\right)}$$

$$\text{Where } P = \frac{F_{\max}}{A_{maincylinder}},$$

$$A_{maincylinder} = \pi d_1^2 / 4$$

$$d_1 = 10 \text{ cm}$$

$$P = \frac{4 \times 5 \times 10^3}{\pi \times 14^2} = 324.8Kpa$$

$$t_h = .433 \times 7\text{cm} \times \sqrt{\left(\frac{324.8Kpa}{125Mpa}\right)}$$

So $t_h = 2\text{cm}$

4.3.4 Design of Main Cylinder

The basic function of the hydraulic cylinder is to convert fluid power into linear machine lift force at that time the cylinder is subjected to internal pressure of the fluid or oil. Since the internal pressure should be high enough to sustain the load then the cylinder must be a heavy-duty cylinder which is expected to be thick to sustain the pressure.

Material selection

Both cast and drawn materials can be selected as cylinder materials for variable height vehicle but the most frequently selected material is called cold drown deep polished low carbon steel with relatively high stress value.

But due to the following reasons we have selected the stainless steel with mechanical properties given below as cylinder material

- Manufacturing feasibility
- Local availability

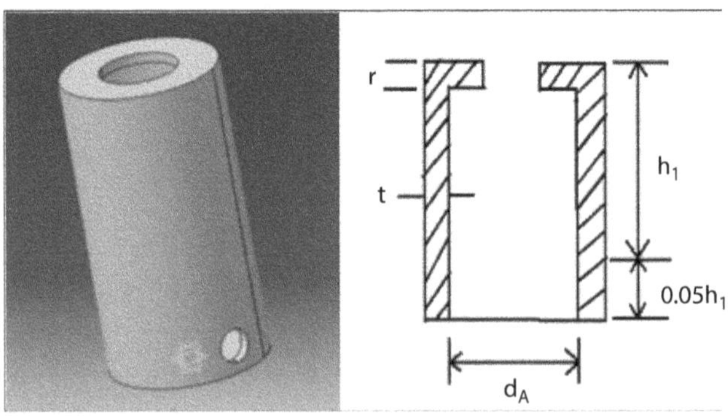

Figure 4.12 Main cylinder.

Material selected	stainless steel
Grade	AISI302
Ultimate tensile stress	σ_u=655Mpa
Yield tensile stress	σ_y=260Mpa
Allowable shear stress	τ_{all}=150Mpa
Modules of elasticity	E=200Gpa
Modules of rigidity	G=77Gpa

Design snalysis

The wall develops both tangential and radial stress with valves which depend upon the radius.

- Assuming that the longitudinal fibber of the cylinder is equivalently strained
- We can analyze the design as follows

Now the pressure and the thickness can be analyzed using Lame's equation as follows.

The cylinder is subjected to both radial and tangential stress $\sigma_r$andσ_t.

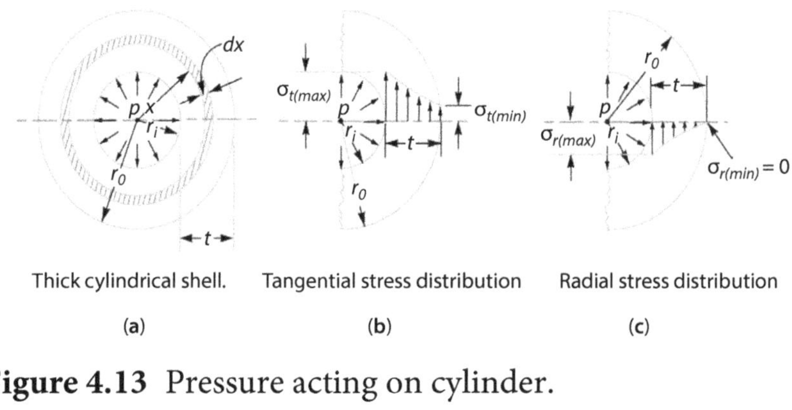

Figure 4.13 Pressure acting on cylinder.

- tangential stress at any radius x

$$\sigma_t = \frac{p_i r_i^2 - p_o r_o^2}{r_o^2 - r_i^2} + \frac{r_i^2 r_o^2}{x^2}\left[\frac{p_i - p_o}{r_i^2 - r_o^2}\right]$$

Where, p_i= inner pressure
r_i=inner radius
r_o=outer radius

- radial stress at any radius x is given as:

$$\sigma_r = \frac{p_i r_i^2 - p_o r_o^2}{r_o^2 - r_i^2} + \frac{r_i^2 r_o^2}{x^2}\left[\frac{p_i - p_o}{r_i^2 - r_o^2}\right]$$

Since there is no external pressure $p_O = 0$. Then,

$$\sigma_t = \frac{p r_i^2}{r_o^2 - r_i^2} - \frac{r_i^2 r_o^2}{x^2}\left[\frac{pi}{r_i^2 - r_o^2}\right]$$

$$= \frac{pi r_i^2}{r_o^2 - r_i^2}\left[1 + \frac{r_o^2}{x^2}\right]$$

σ_t is maximum at x=r_i and the minimum at x=r_o

$$\sigma_{tmax} = \frac{pi[r_i^2 + r_o^2]}{r_o^2 - r_i^2}$$

$$\sigma_{tmin} = \frac{2 pi r_i^2}{r_o^2 - r_i^2}$$

$\sigma_r = \dfrac{pi r_i^2}{r_o^2 - r_i^2}\left[1 + \dfrac{r_o^2}{x^2}\right]$ is maximum at x=r_i and 0 at x=r_o

$$\sigma_{rmax} = -pi \ (Compressive)$$

- Now we can use the maximum strain energy theory to evaluate the cylinder failure, Where Bernie equation is derived to calculate the cylinder thickness.
- $r_o = r_i + t$ and the maximum strain theory we get the following equation called Bernie's equation

$$t = r_i\left[\sqrt{\frac{\sigma_t + (1-\mu)\,p_i}{\sigma_t - (1+\mu)\,p_i}} - 1\right]$$

Where μ = poison's ratio

$$r_i = \frac{d_{maincylinder}}{2} = \frac{d_1}{2}$$

$\sigma_t = \sigma_y/n$ n = factor of safety

- From geometric analysis the internal diameter of the main cylinder was found to be d_1 = 14cm for diameter of the piston rod (d_p=8cm), but during piston design d_p was corrected to 7cm
  Thus the modified main cylinder is;
  d_1 = 7+2(head thickness of piston) +2(clearance between piston head and the cylinder)
  d_1 = 7+2+2c
  = 7+2(1.5) +2(1)
  = 10cm
- Let us take factor of safety n=2.3

$$\text{Then } \sigma_t = \frac{\sigma_y}{n} \sigma_t = \frac{260mpa}{2.3} = 113\text{Mpa}$$

- Oil pressure in cylinder tube can be found as:

Balancing force vertically

$$\Sigma F_y = 0 \qquad p_i * A_1 + ma = 5\text{KN}$$

Assuming the piston is moving upward with uniform velocity the acceleration a=0

$$p_i * A_1 = 5\text{KN Where } A_1 = \pi d_1^2/4$$

Where, d_1- ismain cylinder diameter.

$$P_i = \frac{5\times10^3}{\pi d_1^2/4} = 324.8N/cm^2 = 324.8\,Kpa$$

- This pressure is the same over the walls of the tube
- The thickness of the main cylinder is

$$t = \frac{d_1}{2}\left[\sqrt{\frac{\sigma_t + (1-\mu)p_i}{\sigma_t - (1+\mu)p_i}} - 1\right]$$

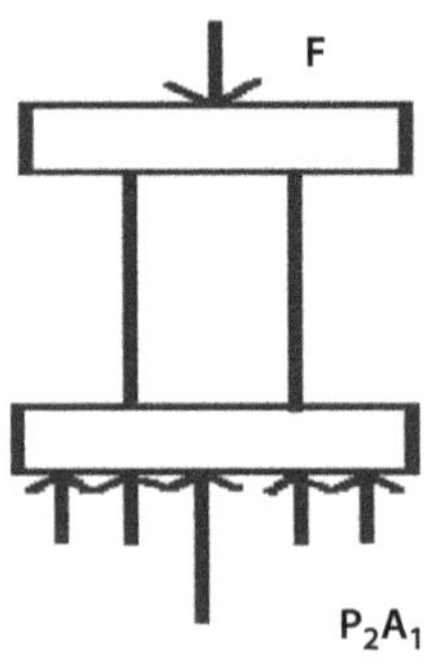

Figure 4.14 Force acting on piston.

μ=Poisson's ratio can be found as follows:

Elasticity and rigidity modulus are related by the following equation :

$$G = E/2(1+\mu)$$

$$2(1+\mu) = E/G$$

$$\mu = E/2G - 1$$

$$\mu = \frac{200Gpa}{2(77Gpa)^{-1}}$$

$$\mu = 0.3$$

Then the thickness t is,

$$t = \frac{14}{2}\left[\sqrt{\frac{113Mpa + (1-0.3)\,32.48}{113Mpa - (1+0.3)\,32.48}} - 1\right]$$

$$t = 9\text{mm}$$

The thickness of the main cylinder can be taken as $t = 9$mm

4.3.5 Design of Reservoir Cylinder

The reservoir must be stable non-reactive material; it is not subjected to any type of stress except when the cover plate is tightened the stress resulting from this tightening action can be neglected since it may not bring significant effect.

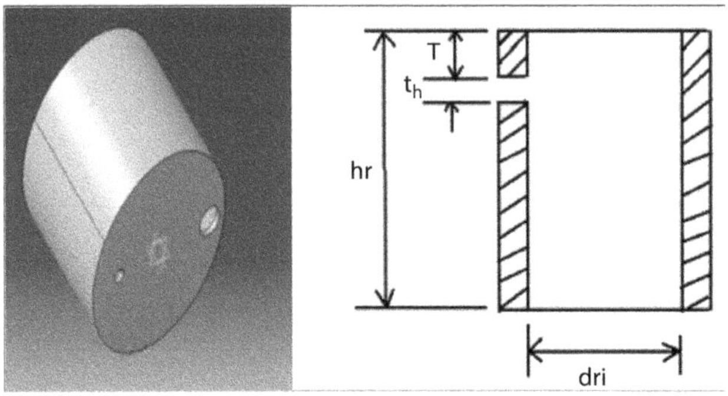

Figure 4.15 Reservoir cylinder.

Material selection

The reservoir only stores the oil and is not subjected to any type of iron used in pipe manufacture and can be selected as previous cylinder material but its inner walls are chromium plated to prevent corrosion and wear. All the oil that fills the main cylinder comes from the reservoir by way of the tube.

Volume of reservoir =volume of the four cylinders + volume of the tube

$$V_R = 4V_{\text{main}} + V_{\text{tube}}$$

From geometric analysis

$$V_{main} = \frac{4 * \pi * d_1^{\,2}}{4} * 27cm$$

$$= \frac{4 * \pi * 14^2}{4} = 16625cm^3$$

$$V_{tube} = L * Area$$

$$V_{tube} = 10m * \frac{\pi * d_{tube}{}^{2}}{4}$$

$$V_{tube} = 10m* \frac{\pi*(0.03)(0.03)}{4}$$

$$V_{tube} = 0.225$$

$$V_R = 4V_{main} + V_{tube}$$

$$V_R = 4(16625) + 0.225$$

$$V_R = 66501.45cm^3$$

$$V_R = \frac{\pi* dr^2 * h}{4} = 66501.45cm^3$$

Let assume that h = 50cm

$$dr = 42 \text{ cm}$$

Let us select thickness of the reservoir t_r = 5 mm

4.3.6 Design of the Pumping Cylinder

The pumping cylinder is small in size when compared to the other cylinder, but is subjected to the same pressure (p = 324.8Kpa).

The analysis is the same with the previous cylinder analysis,

$$d_2 = d_1\sqrt{}R2/20 \text{ where d1=100mm}$$

$$d_2=100\sqrt{}260/20 = 60 \text{ mm}$$

Material for pumping cylinder

Normalize steel of carbon content of 2% yield strength σ_y =250Mpa. This material satisfies the following criteria

- Availability
- Corrosion resistance and mechanical stability
- Manufacturing feasibility

Design analysis

The pump is subjected to both tangential and radial stress

Tangential stress $\sigma_t = \frac{d_2^2 pi}{(d_2^o)^2 - d_2^2}\left(\frac{1+(d_2^o)^2}{(2r)^2}\right)$ where

d_2=internal diameter

d_2^o = outer diameter

r =point of maximum pressure

The maximum pressure at r=d_2/2 thus

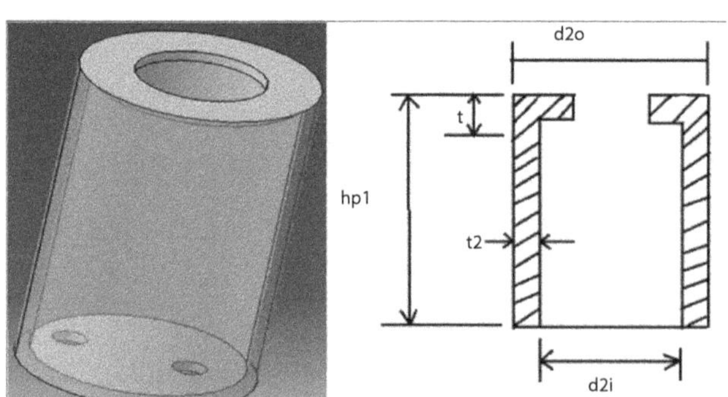

Figure 4.16 Pumping cylinder.

$$\sigma_t = \frac{d_2^2 pi}{(d_2^o)^2 - d_2^2}\left(\frac{d_2^2 + (_2^d o)^2}{d_2^2}\right) = \frac{pi(d_2^2 + (d_2^o)^2)}{-d_2^2 + (d_2^o)^2}$$

The tangential stress σ_t can be

$$\sigma_t = \sigma_{all} = \frac{\sigma_y}{f.s}$$ f.s=2.5(common for steel)

= 250Mpa/2.5 =100Mpa

From geometry analysis d_2=60mm

Thus $$\sigma_t = \frac{pi(d_2^2 + (_2^d o)^2)}{-d_2^2 + (_2^d o)^2}$$

$$100Mpa = \frac{0.32(4761 + (_2^d o)^2)}{(_2^d o)^2 - 4761}$$

d_2^o = 67mm

Thus, thickness of the cylinder is

$$t_2 = \frac{_2^d o - d_2}{2} = \frac{67-60}{2}$$

t_2=3.5mm = Thickness of the pump cylinder

4.3.7 Plunger Design

The plunger is the pump cylinder.

Material selection is the same material as the main piston which can be used as plunger material.

Design analysis

Force on the plunger in compression force due to pressure

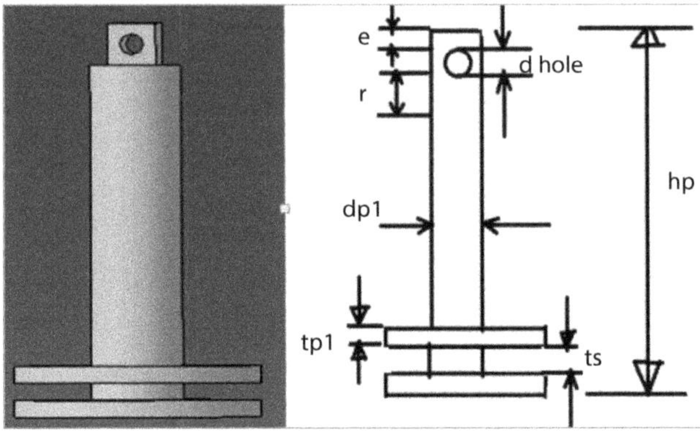

Figure 4.17 Plunger.

- Shear force in the hole
- If the length h_p is large relative to its diameter then the force may cause buckling
- Let us take f.s 2.3

Now to find the plunger lead thickness t_{pl} we use the following equation

$t_{pl} = 0.433 d_2 \sqrt{\frac{p_{\max}}{\sigma_{all}}}$ where t_{pl} = plunger lead thickness

d_2 = bore diameter of cylinder

$p_{\max}$ = 324.8Kpa

σ_{all} = 250Mpa/2.3 = 108.6Mpa

$$Thust_{pl} = 0.433 * 69 \sqrt{\frac{324.8Mpa}{108.6Mpa}}$$

$$t_{pl} = 1.6 \text{ mm}$$

Tearing of the plunger at the hole due to vertical reaction force

$$\sigma_{all} = \sqrt{\frac{R1}{\text{Atear}}}$$

$$Atear = 4\sqrt{\frac{d_{pl}^2 - d_{hole}^2}{4}\ \frac{(dhole + 2l)}{2}}$$

The above equation of the tear area can be derived as follow,

As shown in the figure there is two possible tear areas

$$A = \left(\frac{dhole}{2} + e\right)(2r)$$

$$Buty^2 = \frac{(dpl)^2}{2}\ \frac{(dhole)^2}{2}$$

From Pythagoras tearoom

$$Y = \sqrt{\frac{(dpl)^2}{2}\ \frac{(dhole)^2}{2}}$$

$$A = 2\left(\sqrt{\frac{(dpl)^2}{2}\ \frac{(dhole)^2}{2}}\right)\left(\frac{dhole}{2} + e\right)$$

$$Atear = 2A = 4\left(\sqrt{\frac{(dpl)^2}{2} - \frac{(dhole)^2}{2}}\right)\left(\frac{dhole}{2} + e\right)$$

Considering the crushing of the material diameter of the plunger d_{pl} is given by:

$$\sigma_y = 250\text{Mpa} \quad \sigma_{all} = 109\text{Mpa}$$

$$\sigma_c = \sigma_{all} = \frac{R}{AC} = \frac{4.87 * 1000}{\pi d_{pl}^2 / 4}$$

$$d_{pl} = \sqrt{\frac{4.87 * 1000 * 4}{109 * \pi}} d_{pl} = 40mm$$

$$Then\, A_{tear} = 4\left(\sqrt{\frac{(7.5mm)^2}{4} \frac{(4mm)^2}{4}}\right)\left(\frac{4mm}{2} + e\right)$$

$$A_{tear} = 17\text{mm}^2 + 12\text{emm}^2$$

$$\sigma_{all} = R_2 / A_{tear}$$

$$109Mpa = \frac{4870}{17mm^2 + 12emm^2}$$

$$4870 = 1853 + 1308e$$

$$\text{e} = 2.3 \text{ mm}$$

Height of the piston,

$$hp = hA + e + dhole2 + 2t_{pl} + t_s + r$$

$$= 22.5\text{mm} + 4\text{mm} + 2(1.6) + 1 + 3\text{hp} = 33\text{mm}$$

- The inner seal is located 8mm below the upper surface of the plate a=8mm the thickness of the seal t is to be determined in seal design t_b =15mm geometry analysis.
- Tb only value greater than $2t_s$ but less than $t_{b.}$

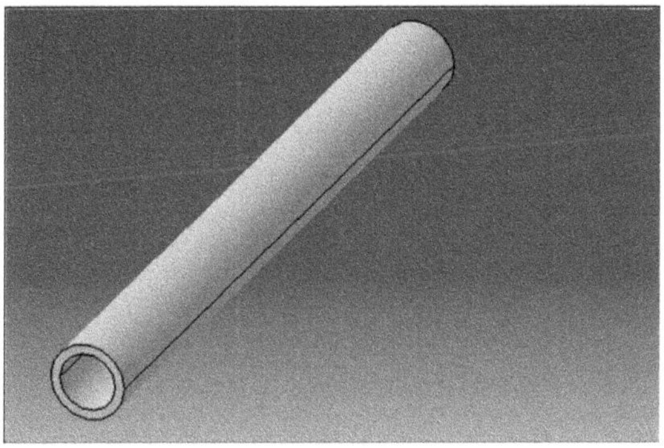

Figure 4.18 Pipe.

4.3.8 Design of Pipe

Inside diameter of pipe $D = 1.13\sqrt{Q/V}$

Thickness of pipe $t = \frac{pd}{2\sigma_t} + c$

Let assume that $Q = 40\ m^3/min$ $V = 2000\ m^3/min$

$$D = 1.13\sqrt{\frac{40}{2000}}$$

$$D = 25\ mm$$

From the weight of the car and diameter of cylinder. We got pressure of 324.8Kpa.

From Table 8.2, we find for steel pipe c=3mm there for thickness of the pipe and $\sigma_t = 40$ Mpa

$$t = \frac{32.48 * 50}{2 * 40} + 3$$

$$t = 3\ mm$$

4.3.9 Design of Spring

Mechanical springs are used in machine to exert force to provide flexibility and to store or absorb energy and I select the helical compressive spring. These are:

- Ease in manufacturing and reliability
- Wide range and constant spring rate

Material selection

We select music wire that has

- Toughness and tensile strength
- Higher stress under cyclic loading

For the material, shear modulus, G = 81.7Gpa
Tension modulus E = 200Gpa
Gage no = 16
Diameter of wire = 2mm
Design shear stress= 895Mpa

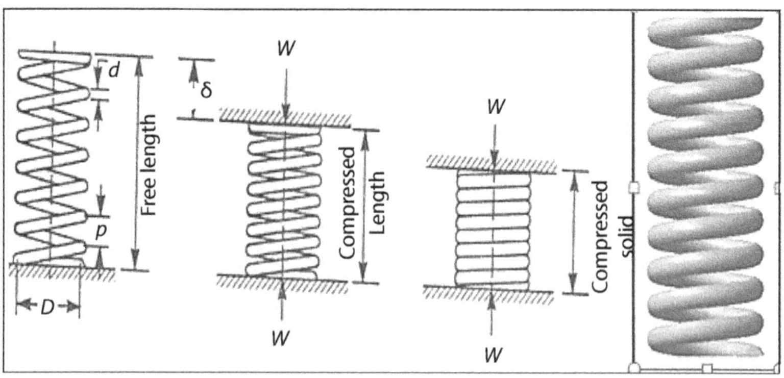

Figure 4.19 Spring.

Table 4.4 Number of active coils specification.

	Squared & ground end	Planed end	Ground end
Na	N-2	N	N-1

To start the design analysis the above standard

And initial specifications are:-

Outside diameter, D_O= 12mm
N_t = total Number of coil =12
Rob oust linearity = 0.15

Where, N_t = coil numbers
N_a = active coil numbers
L_s = spring solid length
L_o = spring free length

The ends are square and ground ended because springs with this end are frequently mounted on a bottom-type seat or on a socket with a depth equal to the height of just a few coil for the purpose of locating the spring.

STEP ONE:

Mean diameter, $Dm = Do - Dw$ = 12mm–2mm = 10mm

Internal diameter, $Di = Dm - Dw$ = 10mm–2mm = 8mm

Spring index, $C = Dm/Dw$, = 100mm/2mm = 5

Which is safe with typical machinery spring having C value from 5-12

Wahl factor,

$$K = \frac{4C-1}{4C-4} + \frac{0.615}{C}$$

$$K = \frac{(4\times 5-1)}{(4\times 5)-4} + \frac{0.615}{5}$$

$$K = 1.3105$$

STEP TWO: [1]

Stress in spring at F = Fo

Where Fo = operating spring force (calculated from given handle force).

$$F = 260\ N$$

As spring is compressed it gets twisted and thus the shear stress may be expressed as:

$$\tau = Tc/J$$

The modified stress equation is given by Wahl as:

$$\tau = \frac{8\times K\times F\times Dm}{\pi(Dw)^3} = \frac{8\times K\times F\times C}{\pi(Dw)^2},$$

$$\text{where } C = \frac{Dm}{Dw}$$

So all the factors are found from the above calculation

$$\tau = \frac{8\times 1.3105\times 260N\times 5}{\pi\times(2mm)^2} = 270Mpa$$

STEP THREE

Deflection at operating force (fo).

$\Theta = TL/GJ$ where θ= angle of twist in radians

T= Torque

L= Wire length

G= Elasticity modulus of the material in shear

J= Wire moment of inertia

Again, for convenience, we will use a different form of the equation in order to calculatethe linear deflection of the spring from the typical design variables of the spring. The resulting equation is,

$$Na = Nt-2$$

$$Na = 12-2$$

$$Na = 10$$

$$\delta = \frac{8\times \mathrm{F}\times \mathrm{Na}\times (\mathrm{Dm})^3}{\mathrm{G}\times (\mathrm{Dw})^4}$$

$$\delta = \frac{8\times \mathrm{F}\times \mathrm{Na}\times (\mathrm{C})^3}{\mathrm{G}\times \mathrm{Dw}}$$

$$\delta = \frac{8\times 260\mathrm{N}\times 10\times 5^3}{81.7\mathrm{Gpa}\times 2\mathrm{mm}}$$

$$\delta = 0.002\mathrm{mm}$$

STEP FOUR

Solid length is the shortest possible length that the spring can have as it is not fully compressed to the solid length during operation.

$$Ls = dw(Nt + 1)$$

$$Ls = 2(12 + 1)$$

$$Ls = 26 \text{ mm}$$

STEP FIVE

Design stress from the average stress service for ASTM- (music wire) for:-

$Dw = 2$ & $\tau = 895$ Mpa & operating stress $\tau o =$ 870 Mpa from the step the design is safe.

STEP SIX

BUCKLING –If Lo > 2.63 DM/α the spring will buckle at operating deflection.

So, to evaluate buckling

$$\text{Lo=Ls+(1+pobust linearity)2}$$

$$\text{Lo=26+(1+0.15)2}$$

$$\text{Lo=28.3}$$

$$\text{Lo}<2.63\ DM/\alpha$$

$$28.3\text{mm} < 52.63\text{mm}$$

Therefore the free length of 28.3is less than 52.63 and buckling is unlikely.

4.3.10 Design of Release Valve

The poppet valve is made tighter to compress the spring and push the ball which pushes fluid out.

This valve is against spring force and pressure force.

Where, t = human force F_s = spring force

F_f = force due to the fluid

Consideration, Shear area = 12.7 mm^2

Tensile strength area = 14.2 mm^2

Material selection

- Course pitch threaded
- Cast iron
- Tensile strength 221Mpa

Design analysis

From above force of spring F_s = 260N.

$\frac{T}{J} = \frac{\tau}{d/2}$ Where, T = human torque

J = shear stress

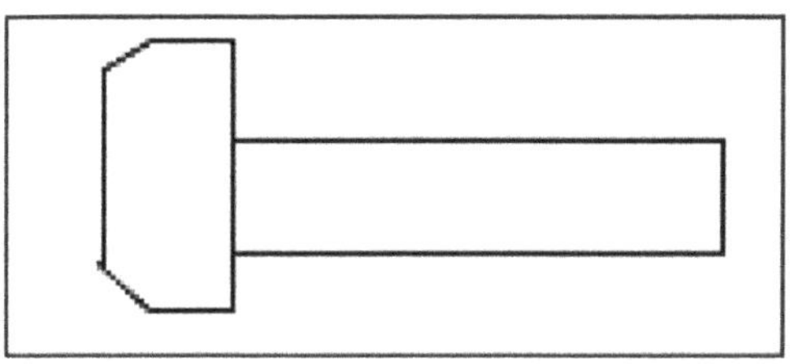

Figure 4.20 Release valve.

τ = shear stress

J = polar moment of inertia

d = valve diameter

$$J = \frac{\pi d^4}{32}\frac{\tau}{T} = \frac{16}{\pi d^3} \text{ Where d = core diameter } \tau = \frac{16T}{\pi d^3}$$

Take diameter to be d = 8 mm

T = f × l = 260N × 450= 117 Nm

D/2 = (F_s× F_f)=T consider F_f = 320

$$\frac{D}{2} = \frac{T}{F_s + F_f} = \frac{4200\,N/mm}{210+320}\frac{D}{2} = 7.93 \quad D = 15.84mm$$

Take the diameter of the head to be 16 mm D = 16mm.

Check strength against compression due to fluid pressure and spring force,

$$\sigma = \frac{Fs + Ff}{area} = \frac{210+320}{14.2} = 37.3239\,N/mm$$

$$F.s = \frac{\sigma_y}{\sigma_{all}} = \frac{221Mpa}{37.33Mpa} F.s = 5.9$$

4.3.11 Design of Pin

Material selection

Malleable cast iron with the following material property can be selected as pin

- ASTM A602

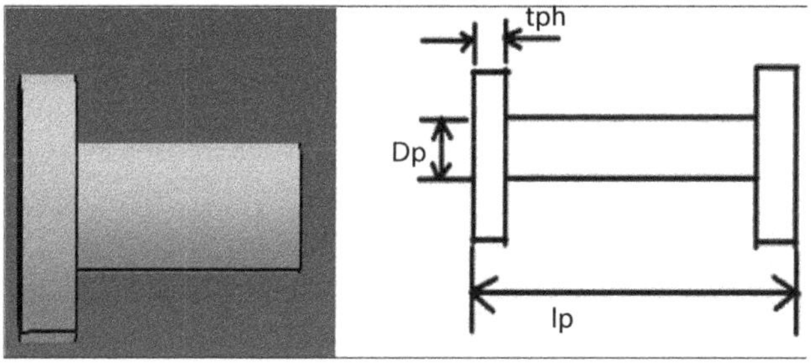

Figure 4.21 Pin.

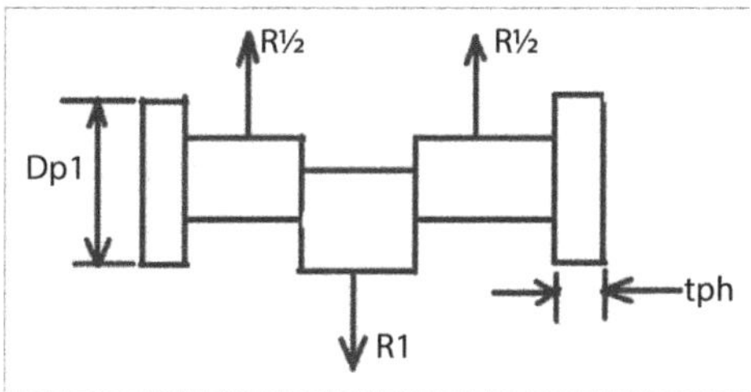

Figure 4.22 Force analysis on pin.

- $\sigma_{TENSILE} = 621$Mpa
- Yield strength $\sigma_y = 483$Mpa
- Let us take factor of safety n = 2

$\tau_{all} = \sigma_y/2n = 150$ MPa

Design analysis

$$Dp1 = dhole + clearance$$

$$= 6mm + 1mm$$

$$= 7mm$$

2 Pin in the plunger socket joint,

$$T_{all2} = \frac{R_2}{2\left(\pi D_{p2}^2/4\right)} \text{where}, D_{p2} = d_{hole2} + clearance$$

$$= 4mm + 1mm$$

$$= 5mm$$

$$T_{all2} = \frac{4.82 * 1000}{2\left(\frac{\pi * 5 * 5}{4}\right)}$$

$$T_{all2} = 130Mpa$$

Since 130<150Mpa there for the design is safe.

4.3.12 Design of Ball

The use of the ball in the hydraulic system is to control the flow of oil out of the cylinder. The ball is pushed by the spring. In order to not go through the spring the ball should have the same or larger diameter than the spring. There is no head to design the ball. This diameter of the ball is taking from standard table. There forms the diameter of the ball is taken to be 12mm.

Material selection

Based on the nature of the load on the ball that is compressive load we select material which is hardest and strongest steel, which is martensitic steel.

4.3.13 Design of Pedal

The pedal is used to transmit force from the human operator to the small piston called plunger. Also used to pressurize the fluid inside actuator and push the required force. In order to design it based on shearing and bending let us put it horizontal and find the minimum shear and bending diagram.

Assumption

In order to get good functioning and get free space we assume in our design the operation is made by inserting its holder, and you can take it from holder while finishing your work.

Material selection

Due to the nature of load applied by the person on the handle we choose gray cast iron (4.5% C, ASTM A-48).

$$\sigma_{ut} = 760\text{Mpa}$$

$$\sigma_{cu} = 665\text{Mpa}$$

$$\tau_u = 240\text{Mpa}$$

Let us use factor of safety for the compensation

$$\text{Let} = \text{F.s} = 1.4$$

$$\text{Load of design} = \text{F.s} \times \text{F}$$

$$F_a = 1.4 \times 20 \text{ kN} = 28 \text{ KN}$$

Where, F_a load of design.

Assumption

x = 30 mm

human effort = 260N
Where θ = angle
leke of handle

with horizontal

FH = human force

Fr = punger force

From assumption FH = 260

Design analysis

$$\sum MA + = 0 \qquad -FH(X+L) + Fr\cos\theta(L)$$

$$FR = \frac{FH(X+L)}{L\cos\theta}$$

$$FR = \frac{260(0.46+0.03)}{0.46\cos 25}$$

$$FR = 4.87KN$$

$$FR_y = FR\cos\theta$$

$$FR_y = 4.87\cos\theta FR_y = 3.94KN$$

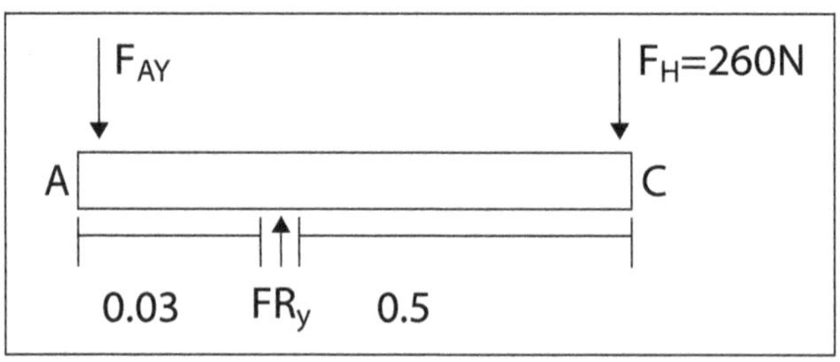

Fig. 4.20 Force analysis on pedal.

$$\sum F_{vertical} = 0$$

$$F_{AY} + 210N - 3709.998 = 0$$

$$F_{AY} = 3.68KN$$

Using section method,

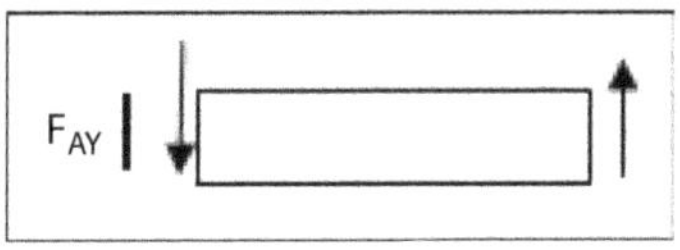

$$V\Sigma Fy = 0$$

$$v = -F_{AY} = 3.68KN$$

From the formula,

$$M_A = -3.68 \times L \quad L = (0, 0.03)$$

$$\text{If } L = 0 \ M_{A=} 0$$

$$\text{If } L = 0.03 \ \ M_{A=} 110.N/mm^2$$

This force is thrash out the distance 0.03m because shearing stress between concentrated load and reaction is constant.

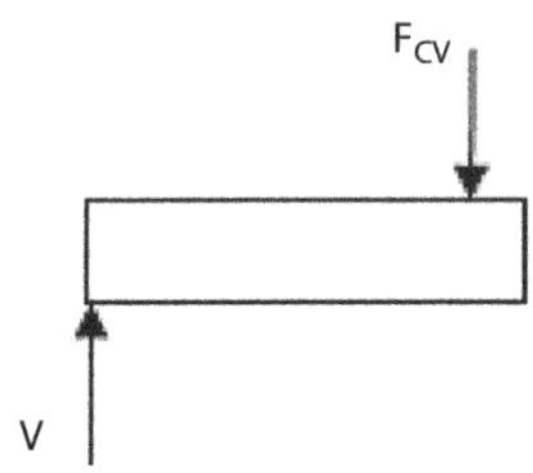

$$\Sigma F_y = 0$$

$$F_{cY} + v = F_{RY}$$

$$v = 3.94 - 3.68$$

$$M_c = 260 \times 1 \qquad 1 = (0,0.5)$$

$$\text{If } L = 0 \quad M_c = 0$$

$$\text{If } L = 0.05 \quad M_A = 130 \text{ N/mm}^2$$

Maximum bending moment is $|M| = |130|$

$$\sigma_{max} = \frac{12|M|}{bh^3} \sigma_{max} = \sigma_{all} = \frac{\sigma_y}{F.s}$$

Let us assume that b = 2h

$$\sigma_{max} = \frac{6|M|}{h^4}$$

Let F.S = 1.25 because it is not big work

$$\sigma_{all} = \frac{170\text{Mpa}}{1.25} = 136\text{Mpa}$$

$$h^4 = \frac{32M}{\sigma_{all}} h = \sqrt[4]{\frac{6 \times 130\text{N.M}}{(136 \times 10^6)}} = 0.0489\,\text{m}$$

$$h = 0.0489\text{m} = 48.9\text{mm} \approx 50\text{mm}$$

$$\text{Then } b = 2h = 2*50 \text{ mm} = 100 \text{ mm}$$

4.3.14 Link Design

The link is connected to the pedal and the base by pin joint. There are two links used to connect the socket and the base.

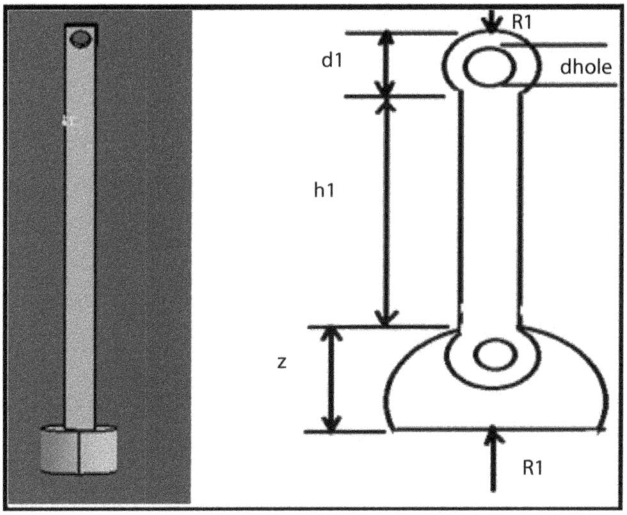

Figure 4.23 Link.

Material selection

- Malleable cast iron
- ASTM A602
- $\sigma_t = 621$ Mpa
- Yield strength $\sigma_t = 483$ Mpa
- Let the least cross section of the link is square of side.the link is subjected to tensile compressive stress due to reaction force $R_1 = 3.68$ KN

$$Thus,\ \sigma_{all} = \frac{R_1/2}{A_{cr}}$$

Where A_{cr} = cross sectional area of the link which must be the least

$$A_{cr} = t_{l1} + t_{l2} \text{ But } t_{l1} = t_{l2}$$

Let us use factor of safety of $\mu = 2\sigma_{all} = \dfrac{483Mpa}{2}$

$$= 241.5 \text{ Mpa}$$

Then $241.5Mpa = \frac{3680/2}{2.5^2} t_l = 2.5mm$

But in practical case the side of the link parallel to the axis of the socket (handle is larger than the other side) $t_{l1} = t_{l2} + 2.5$ mm $= 5$ mm, thus the factor of safety selected above will be higher.

- Considering tearing of the link the shear stress can be found as

$$\tau_{all} = \frac{R_1/2}{A_{tear}}$$

$$A_{tear} = 2(t_{l2})\left(\frac{d_l - d_{holel}}{2}\right) \text{where}, d_{holel} = 6mm$$

$$t_{l2} = 2.5\text{mm}$$

$$\tau_{all} = \frac{\sigma_y}{2F.S} \quad Let\, F.S = 2$$

$$\tau_{all} = \frac{483Mpa}{2*2} = 121Mpa$$

$$\tau_{all} = \frac{R_1/2}{A_{tear}} 121Mpa = \frac{3680N}{2.5mm(d_l - 6mm)}$$

$$d_l - 6mm = \frac{3680N}{121Mpa}$$

$$d_l = 10mm$$

If the link is not to buckle $h_l/k \leq 40$ k=slenderness ratio

$$\text{Or}\, h_l < 40k = \sqrt{\frac{I}{A}} = \sqrt{\frac{t_{l2} * t_{l1}^{3}/12}{t_{l1} * t_{l2}}}$$
$$= 1.44, h_l < 58mm$$

4.3.15 Seal Design

O rings are used in hydraulic cylinder sealing due to their effect need shape rectangular section rings are not suited for reciprocating motion and are only in static application; thus secular shaped seal are more preferable for this; we use o type for both static and dynamic seals.

Material selection

- leather

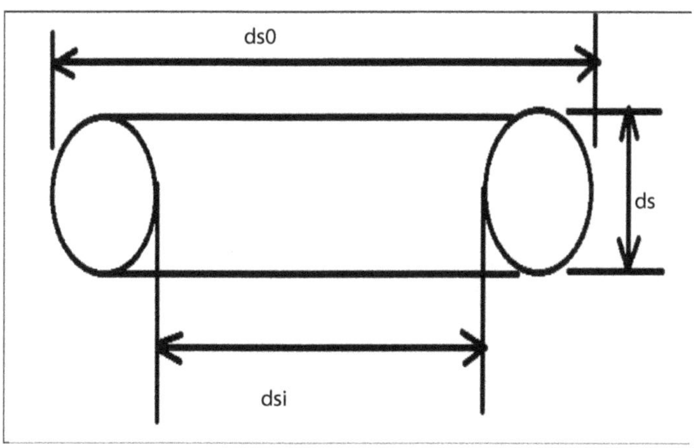

Figure 4.24 Seal.

Reasons to select

- Locally available
- Low cost
- Long life

If D_i is inner diameter of the sealed cylinder then the recommended diameter D_{si} is:

$$D_{si} = (0.95, 0.96, 0.97, \ldots\ldots\ldots\ldots 1)Di - 2d_s$$

Recommended shape of the groove for sealing

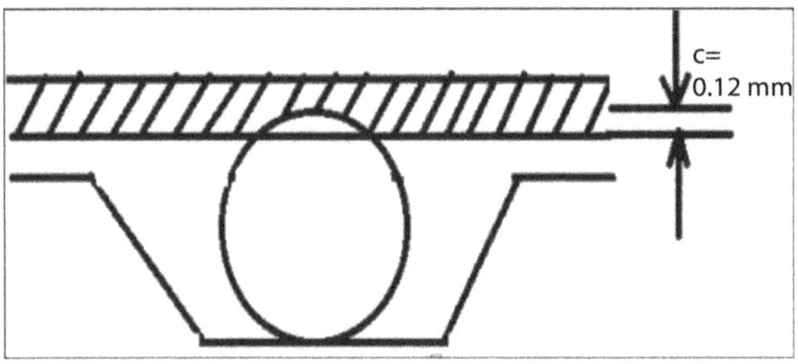

Figure 4.25 Recommended shape of groove for sealing.

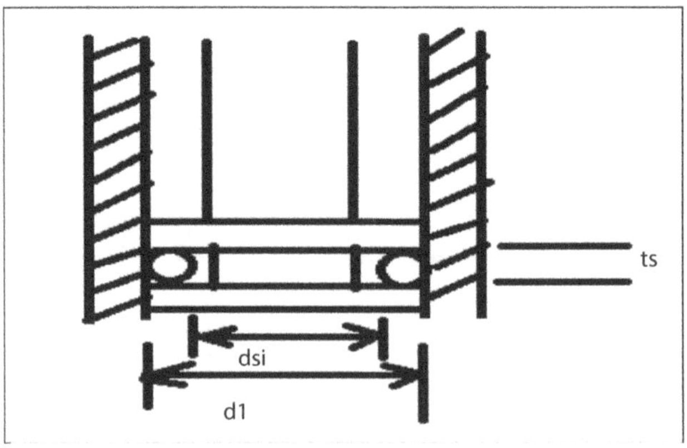

Figure 4.26 Seal in the main cylinder.

The following is least of the seal used in hydraulic system of variable height vehicle design analysis.

Where, ts = cross sectional diameter of the seal and can be found as follows

$$\sigma_{\text{all}} = 0.48 + s$$

- Clearance
- E modules of elasticity
- T_s sealing thickness

Now consider the recommended value of the sealing given above

$$d_{si} = D_i - 2d_s \text{ Where } d_s = t_s - 3\text{mm}$$

$$D_i = d_1$$

In a similar way we can select d_s = 2.5mm for this sealing, thus

$$d_{si} = 0.95D_i - 2d_s \text{ But } D_i = d_1 = 10\text{mm}$$

$$d_{si} = 0.95(10\text{mm}) - 2(2.5\text{mm})$$

$$= 4.5\text{mm}$$

Seal in release valve

$$d_{si} = d_r - 0.12\text{mm} - 2d_s \text{ Let } d_s = 1.5\text{mm}$$

$$d_{si} = 5.12\text{mm} - 3\text{mm}$$

$$d_{si} = 2.12\text{mm}$$

Other seals are also recommended for the manufacturer that each cylinder is perfectly sealed so that energy is saved by reducing the leakage of oil through the clearance.

4.3.16 Testing for Rolling

4.3.16.1 The Longitudinal Location of CG

W -Total car weight,
R_f -Ground reaction at front

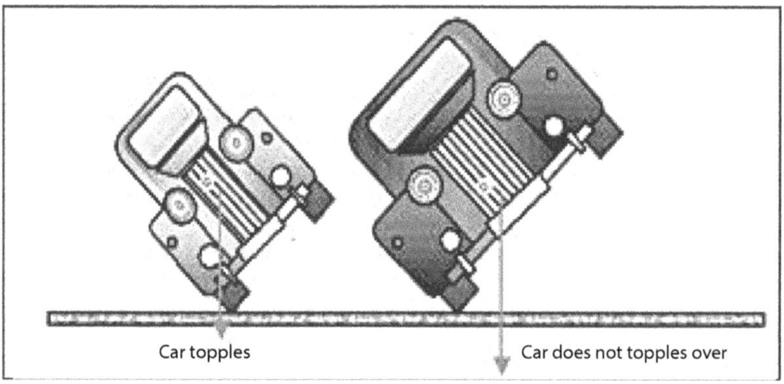

Figure 4.27 Car toppling would be caused due to increasing height of CG or decreasing the width of the track.

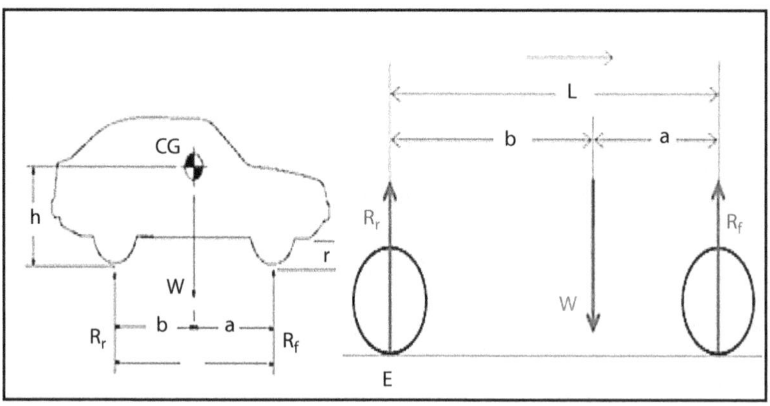

Figure 4.28 CG view (1).

R_r-Ground reaction at rear
R_R-Ground reaction at right wheels
R_L -Ground reaction at left wheels
L -wheel base
T -car track
a –position of CG behind the front axle
b -position of CG in front of the rear axle
x –position of CG from the right wheels
y –position of CG away from the left wheels

The distance between the center of the front and rear wheels (L) can be written as:

$$L = a + b \tag{1}$$

$$a = L - b \tag{2}$$

$$\Sigma F_y = 0$$

$$W - (R_f + R_r) = 0 \tag{3}$$

$$\text{Then } W = (R_f + R_r) \tag{4}$$

Measure the car front axle weight $W_f = R_f$, and the car rear axle weight $W_r = R_r$.

Use equation (4) to find the car weight W.

Take moment around point E.

$$\Sigma M_E = 0$$

$$R_f L - Wb = 0$$

$$R_f L = Wb, \text{ then}$$

$$b = L\,(R_f / W) \tag{5}$$

But, $W = 20000$ N and $R_f = 11500$N

$$b = \frac{4.505m * 11500}{20000}$$

$$b = 2.8\text{m}$$

Where his given as,

Equlibrium equations,

$$M\frac{dV}{dt} = \mu R_F$$

$$R_F + R_R = Mg$$

$$R_R L = Mga + M\frac{dV}{dt}h$$

$$Rf = \frac{Mg(L-a)}{L+\mu h}$$

$$\text{Since } L-a = b Rf = \frac{Mg(b)}{L+\mu h}$$

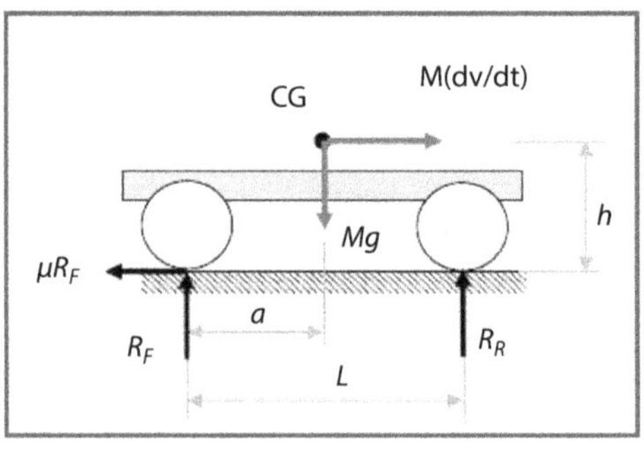

Figure 4.29 CG view (2).

$$11500N = \frac{20000(2.8)}{4.505 + 0.5 * h}$$

$$h = 0.7m$$

But the allowable velocity of vehicle before rising in turning is 30km/hr.

Using this we have to find velocity of vehicle after rising.

h= the location of the CG from the ground before rising.

h_1 = the location of the CG from the ground after rising.

The car can rise up to 20 cm

$$\text{So } h_1 = h + \frac{20}{2}$$

$$h_1 = 0.8m$$

$$M = \frac{mv2}{\text{redias of curviture}} * h$$

$$M = M1$$

$$\frac{mv2}{\text{redias of curviture}} * h = \frac{mv_1^2}{\text{redias of curviture}} * h_1$$

$$V_2 * h = V_1^2 * h_1 V_1^2 = \frac{v_2 * h}{h_1}$$

$$V_1 = \sqrt{\frac{v2 * h}{h_1}} V_1 = \frac{\sqrt{(30^2 * 0.7)}}{0.8}$$

$$V_1 = 22.5 \text{km} / \text{hr}$$

The allowable velocity of vehicle before rising in turning is 30km/hr.

But the allowable velocity of vehicle after rising in turning in order not to roll the vehicle is 22.5km/hr.

Reference

1. **Springs - pdf,Web address:** http://abe.ufl.edu/tburks/Presentations/ABE4171/Springs.pdf Accessed on 18-04-2018.

5

Braking System

5.1 Introduction

The continual adjustment of speed in response to the road condition and traffic density ensures the safety of the driver in a ground transport. This requirement can be effectively met by the braking system, the effective design of which completely ensures the vehicle safety and its reliability. An effective braking system should assure a smooth stop within the shortest possible distance in an emergency situation. Also, the braking should enable the driver to operate the vehicle safely in hilly and muddy terrains without any discomfort to the passengers.

For example, a statistical report published points out that braking problems have been observed in approximately 31.4% of the heavy trucks involved in fatal accidents in the state of Michigan [2]. Based on these requirements and operating conditions, braking systems are commonly classified as service brakes, emergency brakes and parking brakes. The various elements of a brake system are incorporated such that they perform all the aforesaid functions effectively. The braking action can be obtained commonly by following modes (i) By mounting brakes on the respective wheels (ii) By employing brakes on the transmission lines. The former are called foundation brakes whereas the latter are termed retarders. Retarders provide higher braking force on wheels whereas they can only provide minimal braking torque at low speeds, due to which they are used in addition to the foundation brakes. This chapter provides an overview of braking systems used in automobiles, the types of braking systems used in modern automotive systems, and common troubleshooting techniques.

5.2 Background of Brake System

Handbrakes used in cars are used to make them stationary and stop the vehicle while it is moving up a slope. Sudden applications of the brake can put the vehicle out of control and lead to skidding. Most of these brakes are hydraulically operated.

5.3 Classifications of Brake System

1. On the basis of mode of actuation:

    a. Foot brake (also called the main brake) operated by foot pedal.
    b. Handbrake – it is also called parking brake operated by hand.

2. On the basis of mode of operation

    a. Air brakes
    b. Electric brakes
    c. Hydraulic brakes
    d. Mechanical brakes
    e. Vacuum brakes

3. On the basis of action on front or rear wheels

    a. Front-wheel brakes
    b. Rear-wheel brakes

4. On the basis of method of application of braking contact

    a. Externally – contracting brakes
    b. Internally – expanding brakes

5.4 Air or Pneumatic Brake System [3]

The application of the brake by pressing the brake pedal makes the compressed air flow through the delivery port from the supply port. For the front brake circuit, the air in the delivery chamber flows through the pipelines to the frontal brake chamber which is mounted on the front axles of the vehicle. The top bottom supply ports are linked to rear and front reservoirs respectively, while the top and bottom delivery ports are coupled to relay valve and front chambers respectively. Whenever the driver applies force to the plunger, the force transmits to the top piston thereby compressing the rubber plunger spring at the top chamber. If the force applied on the top piston is higher than that of preload and friction, the top piston will move downwards. Meanwhile the exhaust port will be blocked when the top piston makes contact with the top valve, and whenever the force supplied to the top valve can prevail over the preload on the top valve, the inlet port will tend to open. The pressure was kept constant for both the inlet and exhaust ports when closed, thus the air pressure in the top chamber was maintained constant. If the driver reduces the force in the pedal, the balance of the top piston will be broken. The top piston moves up for the force made by compressed air when it is larger than pedal force.

5.4.1 Components of the Typical Air Brake System

1. Compressor and storage reservoirs [1]

Air compressor is the primary energy source for a typical air braking system. Highly pressurized

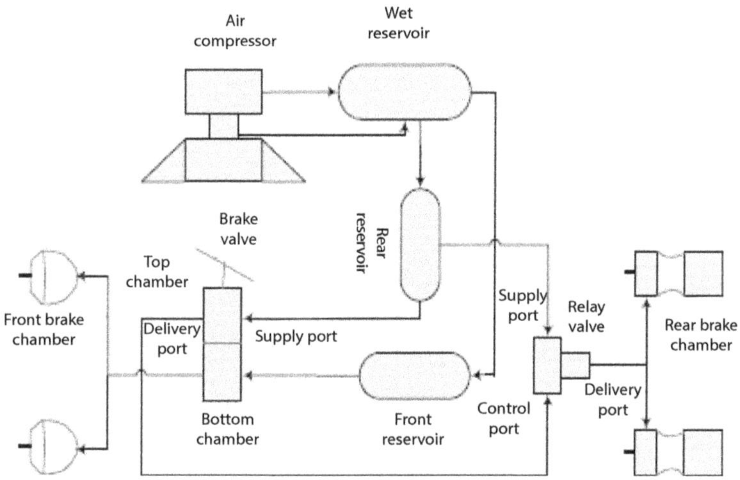

Figure 5.1 Operation of air brake system.

Figure 5.2 Reservoir.

air from the compressor is securely stored in the storage reservoirs or storage tanks where it is allowed to cool down to ambient temperature. The foreign particles and impurities can be expunged from the storage reservoirs by the automatic drain valve.

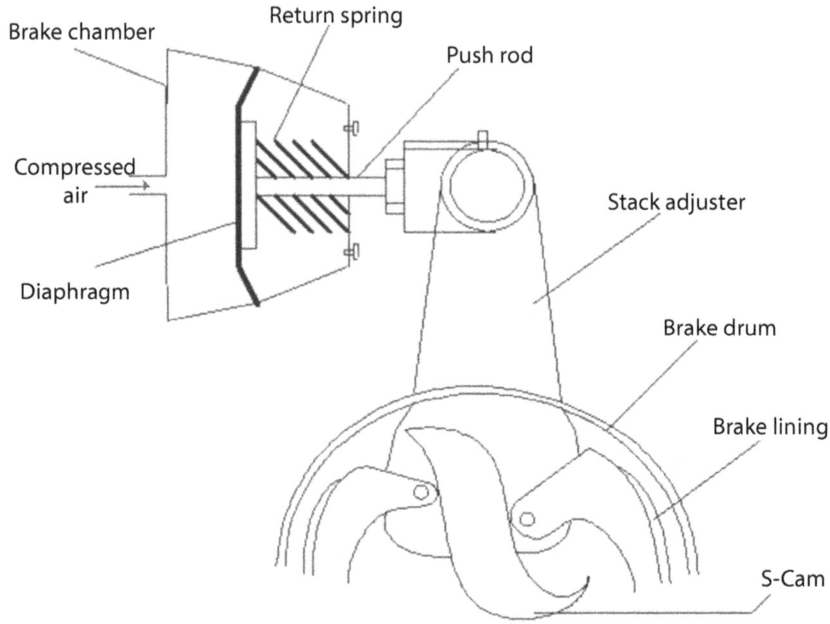

Figure 5.3 Components of air brake system.

2. Treadle valve:

The treadle valve (brake application valve) modulates the amount of air supplied to the brake chambers from the storage reservoirs. The advantage of a dual circuit treadle valve is that partial braking is still possible in the event of failure of one of the two circuits.

3. Quick release valve and relay valve:

A quick release valve is mounted on the front axle of the tractor and the air from the secondary circuit of the treadle valve passes through it en route to the two front brake chambers. It facilitates the quick exhaust of the air from the front brake chambers when the

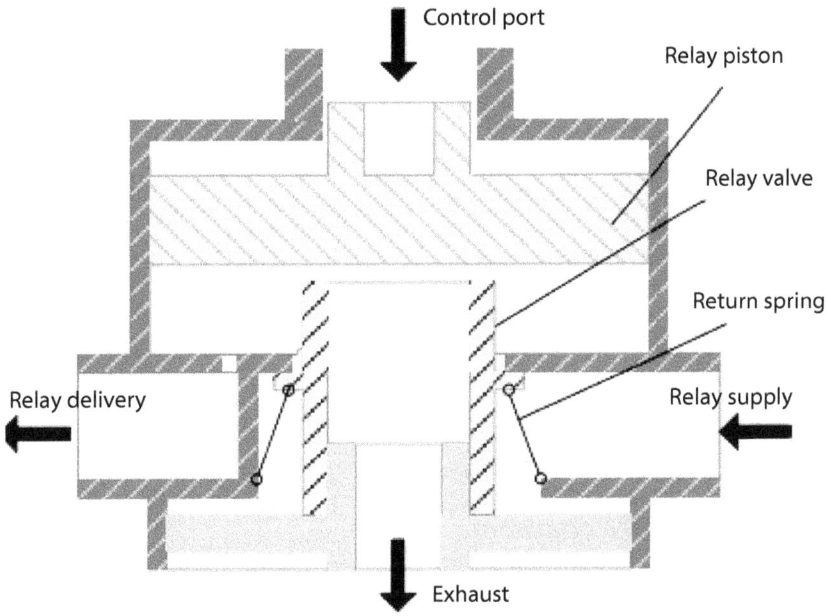

Figure 5.4 Structure of relay valve.

brake pedal is released by the driver. The relay valve also helps in the quick exhaust of air from the rear brake chambers when the brake pedal is released by the driver.

4. Brake chambers

The brake chambers transfer the input energy from the pressurized air into a mechanical energy that shifts the push rod which in turn activates the foundation brakes.

5. Automatic slack adjuster: it converts translation motion of push rod into rotation of cam shaft. The force from rake chamber falls as the stroke of push rod exceeds its highest levels.

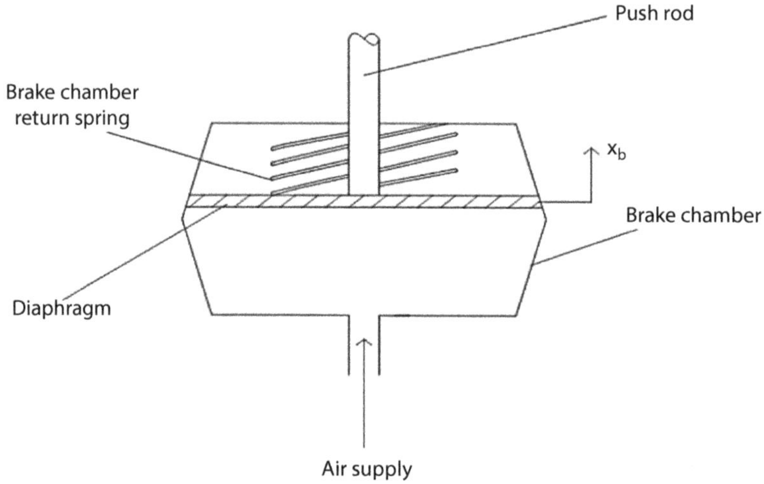

Figure 5.5 Brake chamber.

6. **Foundation brakes**: These units retard the motion of the vehicle and are operated using safety valves which stop building of excessive pressure in storage reservoirs.

7. **Drainage valve:** Removes dirt

8. **Low pressure switch:** Keeps an eye on air pressure

9. **Parking brake control valve:** Used to apply the parking brakes.

10. **Pushrod:** Transmits the force to the slack adjuster.

11. **S-cam:** Moves the brake pads during brake application

12. **Drum-brake:** Consists of two brake–shoes and lining supported on a back plate bolted to the axle-casing

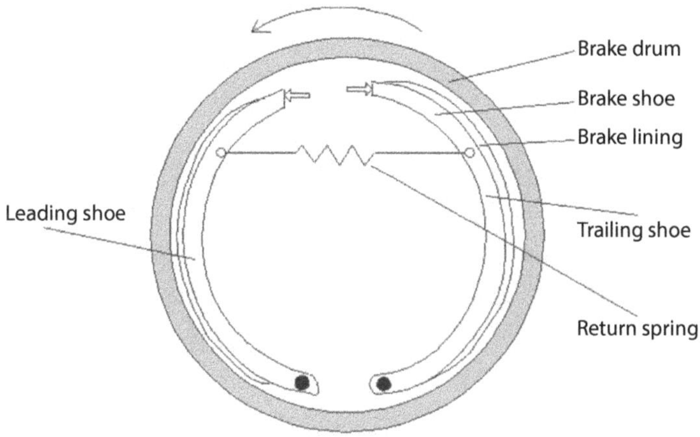

Figure 5.6 Brake drum.

5.4.2 Common Problems in Pneumatic Brake System

- Insufficient brake.
- Brake applies too slowly.
- Brake releases too slowly.
- Brakes do not apply.
- Brake not released.
- Brake grab or erratic brake.
- Air pressure rises above normal.
- Excessive oil or water in the brake system.

5.4.3 Cause of the Problem

- Low air pressure in the brake system.
- Restricted tubing.
- Improper functioning drum/shoe.
- Excessive system leakage.
- Defective governor.
- Compressor passing excessive oil.

- Restricted discharge line.
- Improper alignment

5.4.4 Air Brake System Troubleshooting [4]

1. Dirt is the main cause of premature failure of air valves. So it is safe to blow out the airlines and make sure that it is free of dirt.
2. Whenever the pipe dope is employed on fittings, use it sparingly. This can also lead to failure.
3. It is not advisable to give overtight when installing fittings into a remanufactured valve since it may cause cracking on the valve surface.
4. Make sure that the delivery lines are disconnected from the valve prior to replacing a valve that has air leaking from its exhaust port.

5.4.5 Leakage in Pneumatic Brake System [5]

A graduated Flow Control Valve (FCV) is mounted to get a firm control over the degree of leak in the system. Four turns made on the FCV dial open the valve fully. The main delivery of the treadle valve is connected to one of the brake chambers (front) for all running conditions as depicted in Figure 5.7.

To measure the mass flow rates of the air leaking out from the system, a velocity transducer manufactured

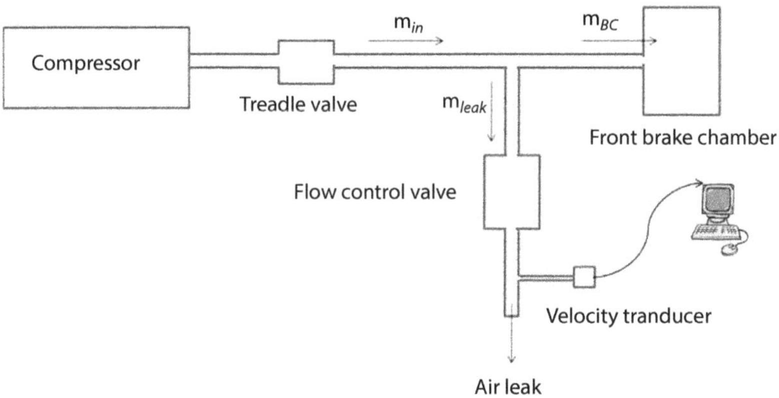

Figure 5.7 Schematic of the leak measurement setup.

by All Sensors Corporation was installed. The sensor was interfaced using the DAQ card and the voltage outputs were recorded during the test runs. The voltage outputs from the sensor were then converted to dynamic pressure using the calibration curve of the sensor. This pressure data was then converted to velocity and in turn to mass flow rate.

5.5 Hydraulic Brake System

A hydraulic brake is an arrangement of the braking mechanism which uses brake fluid, typically containing glycol ethers or diethylene glycol, to transfer pressure from the controlling mechanism to the braking mechanism.

In a hydraulic brake system, when the brake pedal is pressed, a pushrod exerts force on the piston(s) in the master cylinder, causing fluid from the brake fluid reservoir to flow into a pressure chamber through a compensating port. The brake caliper pistons then apply force to the brake pads, pushing them against the spinning rotor, and the friction between the pads and the rotor causes a braking torque to be generated, slowing the vehicle. Alternatively, in a drum brake, the fluid enters a wheel cylinder and presses one or two brake shoes against the inside of the spinning drum. The brake shoes use a similar heat-tolerant friction material to the pads used in disc brakes.

Subsequent release of the brake pedal/lever allows the springs in the master cylinder assembly to return the master piston back into position. The hydraulic braking system is designed as a closed system: unless there is a leak in the system, none of the brake fluid enters

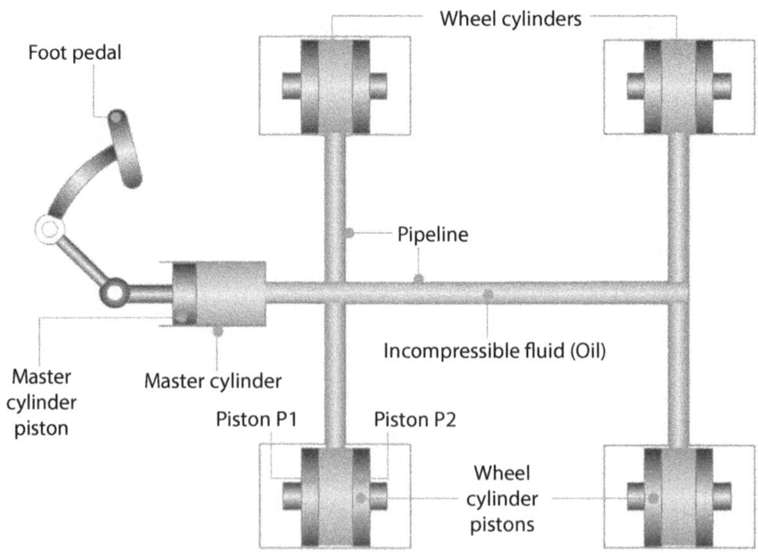

Figure 5.8 Hydraulic brake system.

or leaves it, nor does the fluid get consumed through use. Leakage may happen, however, from cracks in the O-rings or from a puncture in the brake line.

5.5.1 Components of Hydraulic Brake System [6]

The common hydraulic brake arrangement of a commercial vehicle consists of the following:

- Reinforced hydraulic lines
- Brake pedal or lever.
- Pushrod
- Brake caliper assembly
- A master cylinder assembly

5.6 Brake Drum Design

The design or capacity of a brake depends upon the following factors:

1. Pressure between the braking surfaces
2. Coefficient of friction
3. Velocity of the brake drum
4. Projected area
5. Heat dissipation rate

5.6.1 Materials for Brake Lining

The material used for the brake lining should have the following characteristics:

1. High coefficient of friction
2. Low wear rate.
3. High heat resistance.
4. High heat dissipation capacity.
5. Low coefficient of thermal expansion.
6. Mechanical strength.
7. Moisture and oil resistant.

5.6.2 Specification

- max Pressure (p)=1.75 MPa
- Coefficient of friction= 0.2
- diameter of the brake=240mm
- Face width of the brake=30mm
- a(the distance from center to the hinged shoe)=100mm
- Allowable stress of cast iron=276 MPa

e = the perpendicular distance from the actuating force to the hinged point (78+100cos24 = 169.35mm)

- $\theta_1=0^0=0$ rad
- $\theta_2=132^0=2.303$rad

$$Is = \cos\theta 1 - \cos\theta 2$$

$$Is = \cos 0 - \cos 132 = 1.669$$

$$Iss = \left(2\theta 2 - 2\theta 1 + \sin\theta 1 - \sin 2\theta 2\right)/4$$

$$Iss = \frac{2 * 2.303 - 0 + \sin 0 - \sin 264}{4} = 1.4005$$

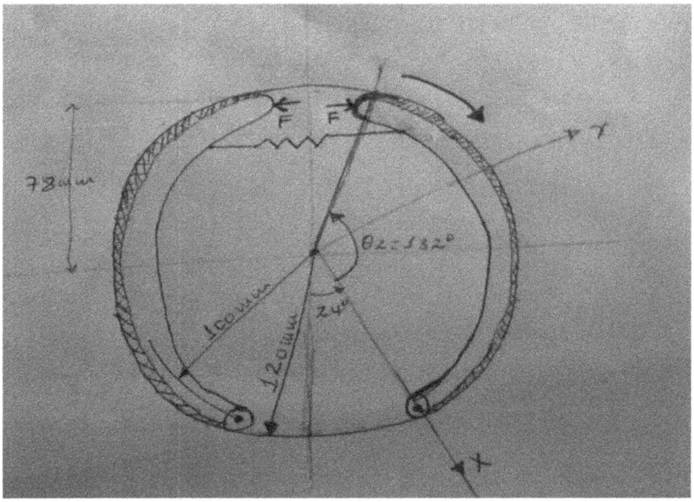

Figure 5.9 Applying force in drum brake.

$$Isc = (\cos 2\theta 1 - \cos 2\theta 2)/4$$

$$Isc = \frac{(\cos 0 - \cos 264)}{4} = 0.2789$$

$$m = \frac{1}{\mu} \cdot \frac{a}{r} \cdot \frac{Iss}{Is}$$

$$m = \frac{1}{0.2} \cdot \frac{100}{120} \cdot \frac{1.4005}{1.669} = 3.496$$

$$n = 1 - \frac{a}{r} \cdot \frac{Isc}{Is}$$

$$n = 1 - \frac{100}{120} \cdot \frac{0.278}{1.669} = 0.867$$

Where, n &m are constant dimensionless characteristics of the shoe.

- Then we are going to identify the maximum pressure of brake sides

Left side

- $\Delta = -1$ because the brake shoe is internal
- $\delta = +1$ because in the left side brake is trailing shoe
- The actuating moment at the bottom of the hinged(M)

$$M_L = F * e$$

- e=the perpendicular distance from the actuating force to the hinged point

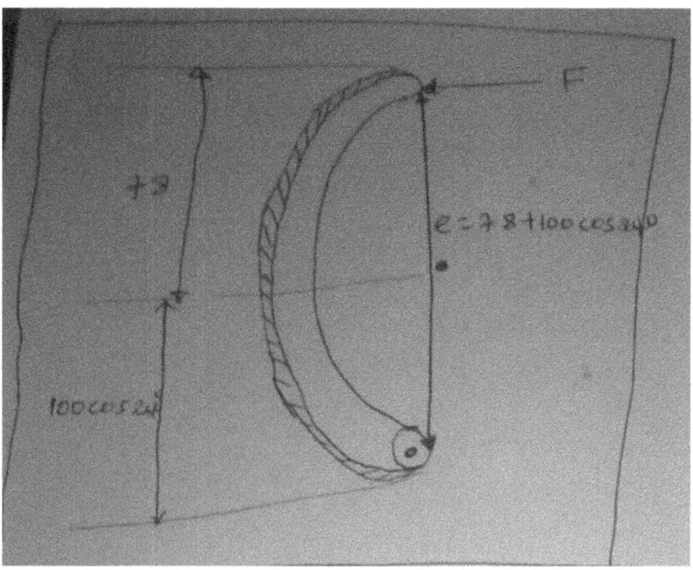

Figure 5.10 Force in the left side.

e=78+100cos24=169.35mm

Let the applied force is 1N

M_L=F*e=1N*169.35mm = 169.35N

$$No = \frac{M_L}{\mu rls(m - \delta\Delta n)}$$

$$No = \frac{169.35Nmm}{0.2 * 120mm * 1.669(3.496 + 0.8607)}$$

$$No = 0.9704N$$

Where, No is pressure parameter.

- The maximum pressure in the left shoe

$$P\max = \frac{No}{w * r}$$

$$P\max = \frac{0.9704N}{30mm * 120mm}$$

$$P\max = 0.26956kpa$$

Next we will calculate the maximum pressure on the right shoe.

Right shoe

- Δ= −1 because the brake shoe is internal
- δ= −1 because in the left side brake is leading shoe

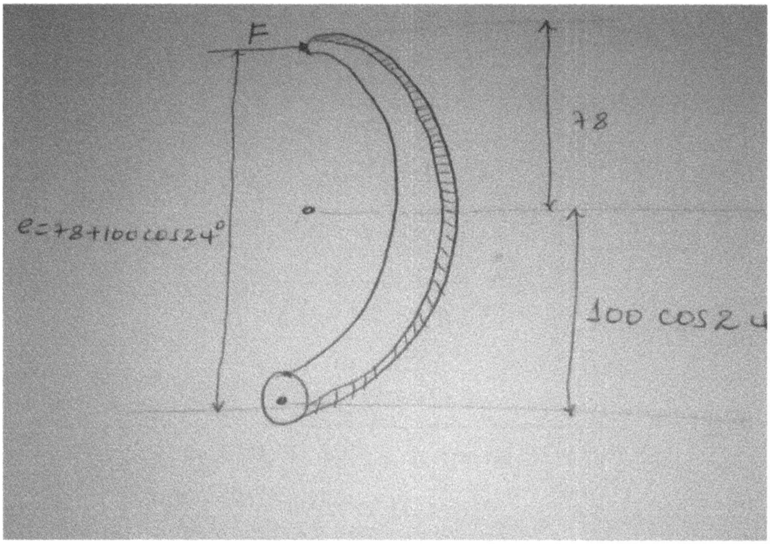

Figure 5.11 Apply force in the right shoe.

- The actuating moment at the bottom of the hinged(M)

  Mr=F*e=1N*169.35mm=169.35Nmm

  e is the same with the left shoe

$$No = \frac{Mr}{\mu r Is(m - \delta\Delta n)}$$

$$No = \frac{169.35Nmm}{0.2 * 120mm * 1.669(3.496 - 0.8607)}$$

$$No = 1.584N$$

The maximum pressure in the right shoe

$$P\max = \frac{No}{w * r}$$

$$P\max = \frac{1.584N}{30mm * 120mm}$$

$$P\max = 0.4401kpa$$

Then the maximum pressure required for the 1N applied force is in the right side Pmax=0.4401kpa.

This pressure is less than the maximum limit pressure (1.75Mpa); therefore our design is safe.

Next we will require the applied force formed by the Pmax

1N = 0.4401kPa

F = 1.75 MPa

Then by cross multiplication the applied force becomes

$$F = \frac{1N * 1.75Mpa}{0.4401kpa}$$

$$F = 3976.36N$$

Over all braking torque

$$T = Tr + Tl$$

For left shoe

$$Tl = \mu * No * r * Is$$

$$Tl = 0.2 * 3858.66 * 120 * 1.669 = 154.56Nm$$

Where, No in the maximum Force applied

$$No = 0.9704*3976.36\text{N} = 3858.66\text{N}$$

Right shoe

$$Tr = \mu * No * r * Is$$

$$Tr = 0.2 * 6300 * 120 * 1.669 = 252.35N$$

$$No = 1.58437 * 3976.36N = 6300N$$

$$T = Tr + Tl = 252.35 + 154.56 = 406.9N$$

Stress on the brake drum

$$\delta h = P\max D / 2w$$

$$\delta h = \frac{1.75Mpa * 240mm}{2 * 30mm} = 7Mpa$$

- D=diameter of the drake drum
- W=face width of the brake drum
- δh=stress

5.6 Result and Conclusion

We try to calculate the braking force, braking torque as well as the stress in the drum brake of the pneumatic brake system at a pressure of 1.75Mpa,1.5Mpa and 1Mpa. Then the results are listed as follows:

Pressure	1.75Mpa	1.5Mpa	1Mpa
Force	3976.36N	3408.316N	2272.2N
Torque	406.9Nm	348.78Nm	232.52Nm
Stress	7Mpa	6Mpa	4Mpa

References

1. **A diagnostic System for Air Brakes in Commercial Vehicles, Web address:** http://www.academia.edu/34233797/A_Diagnostic_System_for_Air_Brakes_in_Commercial_Vehicles Accessed on 18-04-2018.
2. **S. F. Williams, R. R. Knipling, *Automatic slack adjusters for heavy vehicle air brake systems*, Nat. Highway Traffic Safety Administration**
3. Modeling and Simulation Vehicle Air Brake System, Web address: http://www.academia.edu/34233797/A_Diagnostic_System_for_Air_Brakes_in_Commercial_Vehicles Accessed on 18-04-2018.
4. **Air Brake System Troubleshooting, Web Address:** https://www.suspensionspecialists.com/techinfo/

Bepco_Air_Brake_Trbl_Chart.pdf Accessed on 18-04-2018.

5. **Development of a model for an air brake system with leaks, Web Address:** https://pdfs.semanticscholar.org/0922/42de12e0afb3c3e81c41886a036793f9ad50.pdf Accessed on 18-04-2018.
6. **Hydraulic Brake, Web Address:** https://en.wikipedia.org/wiki/Hydraulic_brake Accessed on 18-04-2018.

6

Steering System

6.1 Introduction

Steering is the collection of components, linkages, etc., which allows any vehicle (car, motorcycle, bicycle) to follow the desired course. The collection of components:

- In a car: ensure that wheels are pointing in the desired direction of motion.
- Convert rotary motion of the steering to the angular turn of the wheel.
- Mechanical advantage is used in this case.
- The joints and the links should be adjusted with precision.
- Smallest error can be dangerous.

- No transfer the shocks.
- Minimize the wear.

6.2 Parts of Steering System

The major components are,

1. Pipes and hoses
2. Steering wheel
3. Main shaft
4. Reservoir
5. Pump
6. Steering gear box

6.3 Ackerman's Steering Mechanism

The Ackerman principle is that during turning of I-center of all wheels meet at a point, then the vehicle

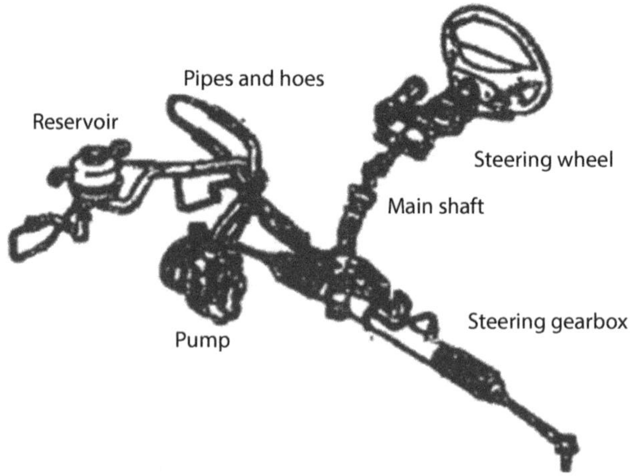

Figure 6.1 Parts of a steering system.

will take a turn at that point which results in pure rolling of the vehicle.

Ackerman condition for two-wheel steering:

Expressed as;

$$\cot \delta_o - \cot \delta_i = \frac{B}{L} \tag{6.1}$$

Where,

δo = Angle of outer wheel
δi = Angle of inner wheel
W = Track width
B = Distance between left and right king-pin centerline
L = Wheel base

Here, Ackerman condition is satisfied when I-centers of front wheels meet at a point on the rear axis of the vehicle which is the turning point of the vehicle.

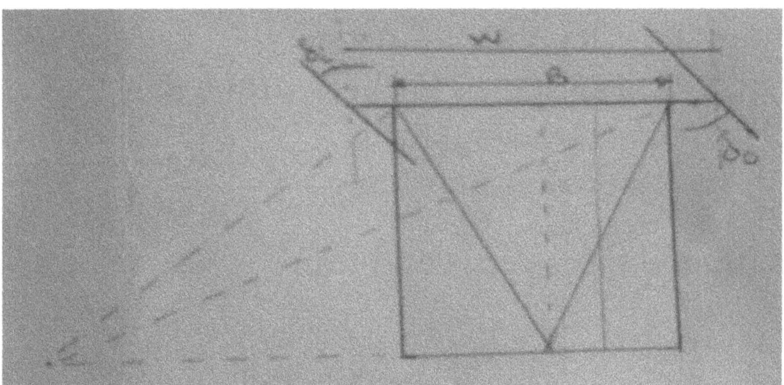

Figure 6.2 Ackerman steering mechanism.

Rack and pinion geometry

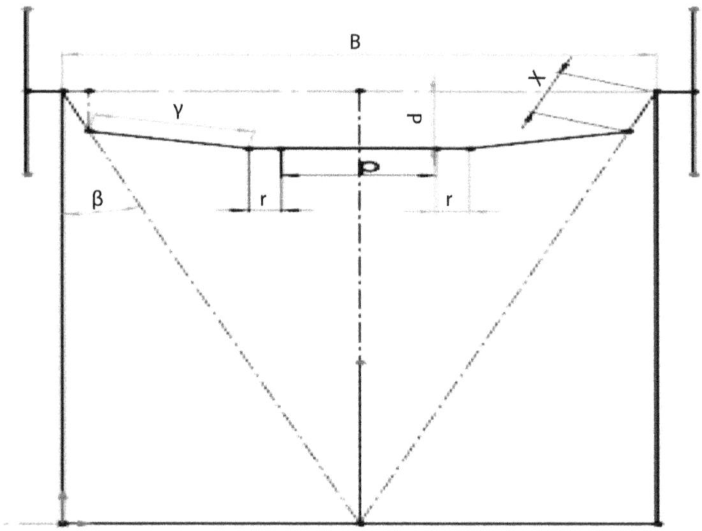

Where,

x = Length of steering arm
y = Length of tie rod
p = Length of rack casing
p+2r = rack ball joint center to center length
q = Rack travel distance
d = Distance between front axis and rack center axis
β = Ackerman angle

Equation: For toe zero condition

$$y^2 = \left[B - \frac{P+2r}{2} - x\sin\beta \right]^2 + \left[d - x\cos\beta \right]^2 \quad (6.2)$$

Equation: From inner wheel geometry

$$y^2 = \left[\frac{B}{2} - \left(\frac{p}{2} + r - q\right) - x\sin(\delta_i + \beta)\right]^2 + \left[d - x\cos(\delta_i + \beta)\right]^2 \quad (6.3)$$

Equation: From outer wheel geometry

$$y^2 = \left[\frac{B}{2} - \left(\frac{p}{2} + r + q\right) + x\sin(\delta_o - \beta)\right]^2 + \left[d - x\cos(\delta_o - \beta)\right]^2 \quad (6.4)$$

- For given rack and pinion, value of **p** and **r** is known.
- Value of **B** is fixed by track width of the vehicle and distance between wheel center andkingpin center.
- Value of **β** is fixed by value of **B** and wheel-base **L**.

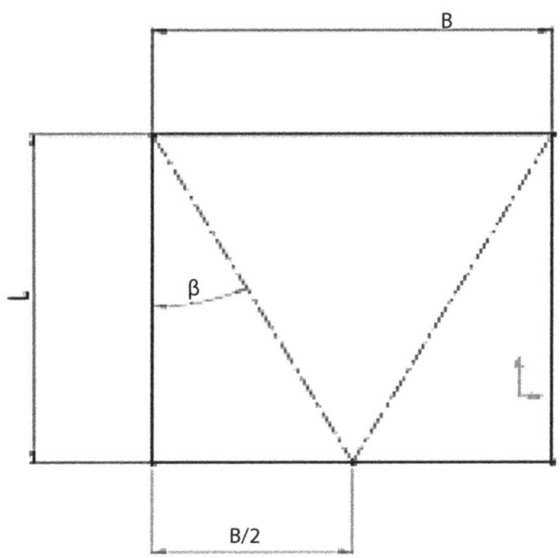

from this, $\beta = \tan^{-1}\left(\frac{2L}{B}\right)$

Angle δ_i is the value of certain inner wheel angle at which we want to satisfy the ackerman principle in order to calculate outer wheel angle δ_o using this equation

$$\delta_o = \cot^{-1}\left(\cot\delta_i + \frac{B}{L}\right) \tag{6.5}$$

- Now, we have 4 unknowns: **x, y, d, q** and 3 equations so any one variable we can fix either according to restriction if any or as per our comfort. After fixing any one variable, we can calculate value of the other three variables by solving these three equations.
- Thus, we have values of all **x, y, d, q**- steering geometry parameters and for this we will get perfect Ackerman condition when inner wheel is at angle **δI** and therefore outer wheel is at angle **δo.**

Rack travel q for any particular inner wheel angle δi:

$$q = x * \sin(\delta_i + \beta) - A + \left[y^2 - \left(d - x * \cos(\delta_i + \beta)\right)^2\right]^{0.5}$$

Where $A = \frac{B}{2} - \left(\frac{p}{2} + r\right)$

Calculation of actual outer wheel angle δ o for any particular inner wheel angle δi:

- $$y^2 = \left[\frac{B}{2} - \left(\frac{p}{2} + r + q\right) - x sin(\delta_o - \beta)\right]^2 + \left[d - x cos(\delta_o - \beta\right]^2$$

- $c = (A - q)\, sin\gamma - d cos\gamma$

Where, $c = \dfrac{\left[y^2 - d^2 - x^2 - (A-q)^2\right]}{2x}$, $\gamma = \delta_o - \beta$

- $c = (A - q)\, sin\gamma - d(1 - \sin^2\gamma)^{0.5}$

$$\left[(A-q)^2 + d^2\right] * \sin^2\gamma$$
$$-2c(A - q)\, sin\gamma) + c^2 - d^2 = 0$$

$$sin\gamma = c * (A - q) + \begin{bmatrix} c^2 * (A-q)^2 - \left\{(A-q)^2 + d^2\right\} \\ * \left(c^2 - d^2\right) \end{bmatrix}^{0.5} = k$$

$$\gamma = \sin^{-1}(k) \text{ but } \gamma = \delta_o - \beta$$

$$\delta_o = \sin^{-1}(k) + \beta \qquad (6.6)$$

To see deviation of designed steering geometry from the perfect Ackerman geometry (Geometry in which at every point of Ackerman principle is satisfied), plot graph for two curves:

1. Outer wheel angle δo as per designed steering geometry v/s inner wheel angle δi.
2. Outer wheel angle δo as per perfect Ackerman steering geometry v/s inner wheel angle δi.
3. Now design steering geometries for different values (iterations) of inner wheel angle δi at which Ackerman condition is satisfied and select such geometry for which optimum deviation from perfect Ackerman geometry as well as steering effort are achieved.

Solved example:

1. Wheelbase L = 1.524 m
2. Track width W = 1.27 m
3. B = 1.137 m
4. β = 20.457 deg
5. For Tata Nano rack p = 0.273 m and r = 0.0635 m
6. x = 0.0753 m is fixed in this example due to restriction in length of steering arm because of knuckle design. Now, here suppose we want to achieve Ackerman condition when inner wheel angle **δi = 40 deg** and therefore as per Ackerman principle **δo = 27.296 deg** for given data.

$$y^2 = [0.3685 - 0.3495x]^2 + [d - 0.9369x]^2,$$

$$y^2 = [(0.3685 + q) - 0.87x]^2 + [d - 0.4931x]^2,$$

$$y^2 = [0.3685 - q + 0.1191x]^2 + [d - 0.9929x]^2$$

By solving these three equations, we will get values **Y, d** and **q**

By applying and solving three equations of mathematical model for any vehicle, rack and pinion Ackerman steering geometry for any vehicle can be designed.

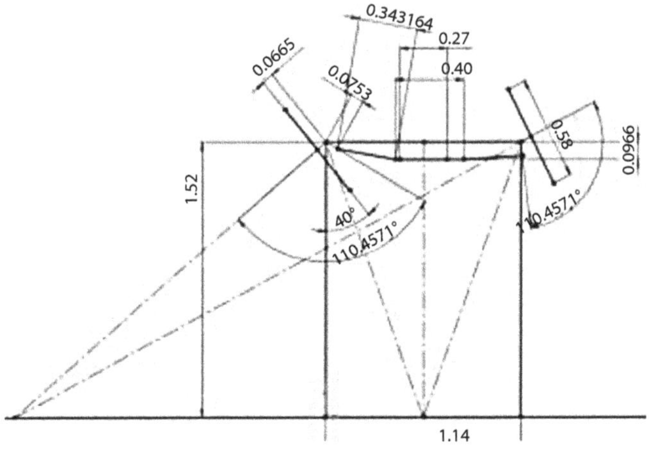

(Travel of rack when inner wheel angle is 40 deg).

$$Y = 0.3431 \text{ m}, d = 0.0966 \text{ m}.$$

Travel of rack q = 0.0349 m (when inner wheel angle δi is 40 deg).

Now, for different inner wheel angle get value of outer wheel angle as per ideal Ackerman as well as actual steering geometry by using (5) and (6) respectively and then plot the graph for outer wheel angle (for ideal Ackerman and for actual geometry) v/s inner wheel angle.

Here graph is as shown below.

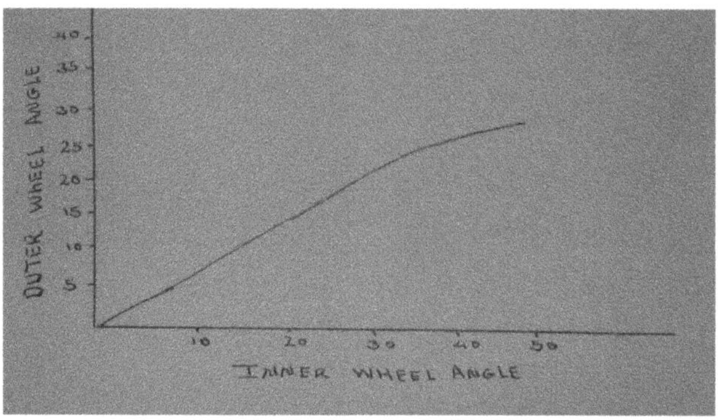

6.4 Davis Steering Mechanism [1]

The Davis steering mechanism shown in Figure 6.3 is an exact steering mechanism which consists of slotted links AM and BH attached to front wheel axle. These slotted links turns on the pivots A and B respectively. With the help of sliding members at P and Q, the rod CD is restricted to move along its length. These constraints are linked to the slotted link BH and AM by turning and sliding pairs at each

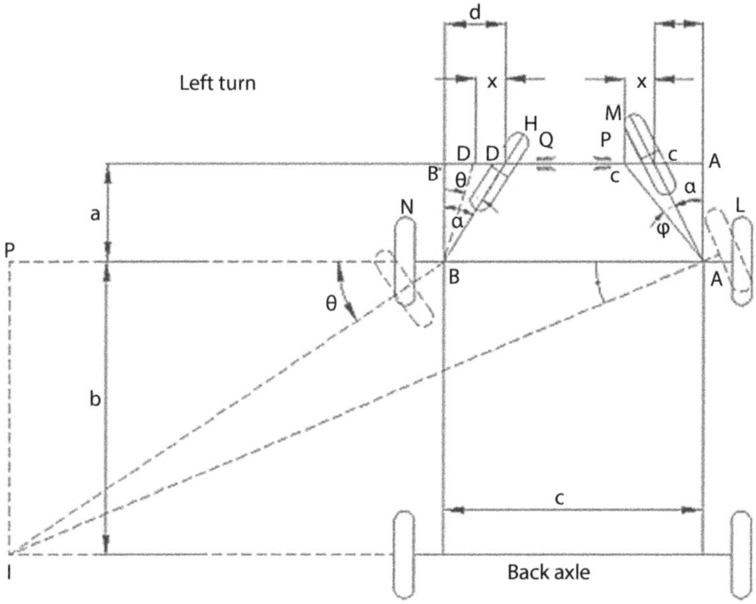

Figure 6.3 Davis steering mechanism.

end. By moving CD to either sides of its nominal position, the steering can be affected. The position of CD for left turn is shown by C′D′.

Let a = Vertical distance between AB and CD,
b = Wheel base,
d = Horizontal distance between AC and BD,
c = Distance between the pivots A and B of the front axle
x = Distance moved by AC to $AC' = CC' = DD'$, and
α = Angle of inclination of the links AC and BD, to the vertical.

From triangle $AA'C'$,

$$\tan(\alpha + \phi) = \frac{A'C'}{AA'} = \frac{d + x}{a}$$

From triangle $AA'C$,

$$\tan \alpha = \frac{A'C}{AA'} = \frac{d}{a}$$

From triangle $BB'D'$,

$$\tan(\alpha - \theta) = \frac{B'D'}{BB'} = \frac{d - x}{a}$$

We know that $\tan(\alpha + \phi) = \dfrac{\tan \alpha + \tan \phi}{1 - \tan \alpha \cdot \tan \phi}$

or $$\frac{d + x}{a} = \frac{d / a + \tan \phi}{1 - d / a \times \tan \phi} = \frac{d + a \tan \phi}{a - d \tan \phi}$$

$$\tan \phi = \frac{ax}{a^2 + d^2 + dx}$$

Similarly, from $\tan(\alpha - \theta) = \dfrac{d - x}{a}$, we get

$$\tan \theta = \frac{ax}{a^2 + d^2 - dx}$$

We know that for correct steering,

$$\cot \phi - \cot \theta = \frac{c}{b} \text{ or } \frac{1}{\tan \phi} - \frac{1}{\tan \theta} = \frac{c}{b}$$

$$\therefore \; 2 \tan \alpha = \frac{c}{b} \quad \text{or} \quad \tan \alpha = \frac{c}{2b}$$

6.5 Power Steering

The basic aim of power steering is to minimize human steering effort and this is achieved by a closed loop hydraulic assisted rack and pinion type steering system. The other aim is to design the rack and pinion gears to smooth out steering by using a helical gear and light weight components so that the steering system provides control over direction of travel, good maneuverability, smooth recovery from turns and minimum transmission of road shocks.

6.5.1 Hydraulic Power Assisted Steering – HPAS System [2]

HPAS system is one of the most widely used steering systems owing to its enhanced sensitivity and smooth operation. The method of utilizing oil under high pressure to boost the servo is advantageous in terms of weight and cost reduction. With closed loop operation shown in Figure 6.5, the inputs like steering wheel angle and torque will be transferred to hydraulic power steering unit which houses valve, steering and hydraulic system which magnifies the turning effect of the respective axles.

6.5.1.1 Static Characteristic of the Hydraulic Power Steering System[2]

Design of a HPAS system has to take into account trade-off between road feel and assistance under different driving conditions. This relationship is given in

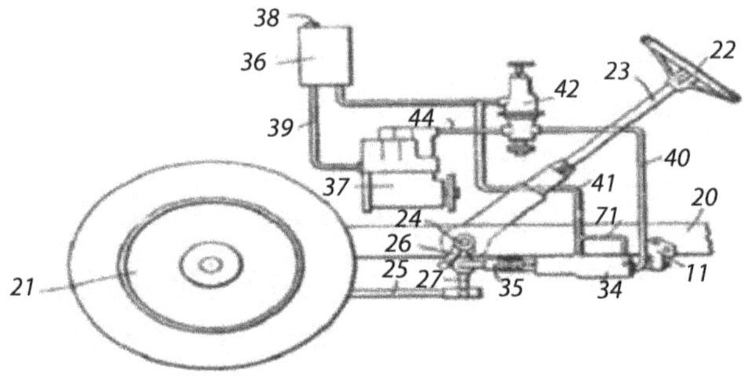

Figure 6.4 Old power steering.

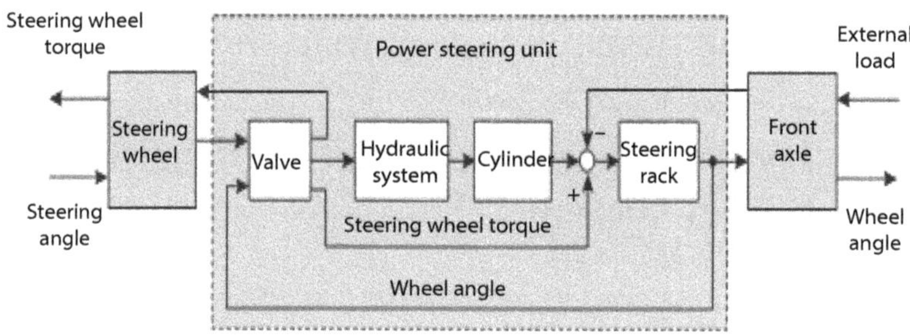

Figure 6.5 The power steering system.

Figure 6.6. Figure 6.7, shows the relationship between wheel torque and load pressure. Figure 6.8, depicts the relationship between assisting and manual forces as a factor of the total generated force (Equation 6.1). From the figure, it is evident that at low torque, the manual force is high to ensure good road feel.

$$F_{tot} = F_{manual} + F_{assist} \qquad \text{Equ. 6.1}$$

$$F_M = \frac{F_{manual}}{F_{total}}$$

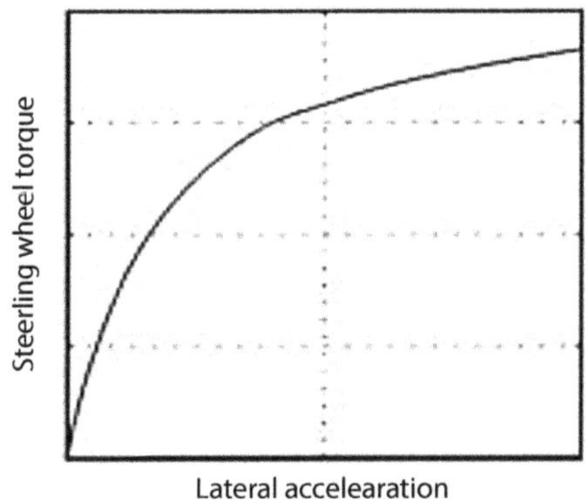

Figure 6.6 Steering wheel torque.

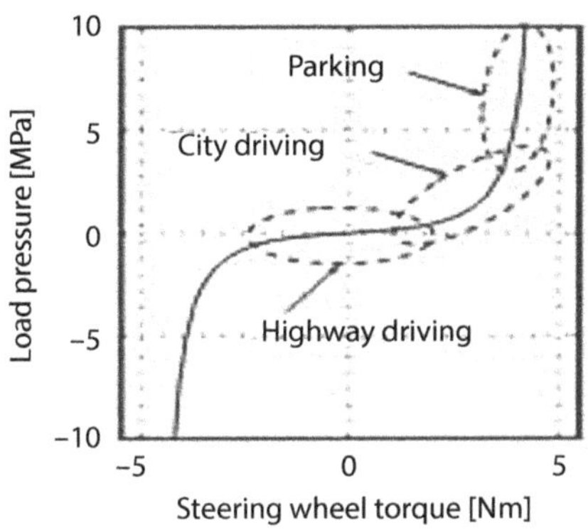

Figure 6.7 Boost curve width.

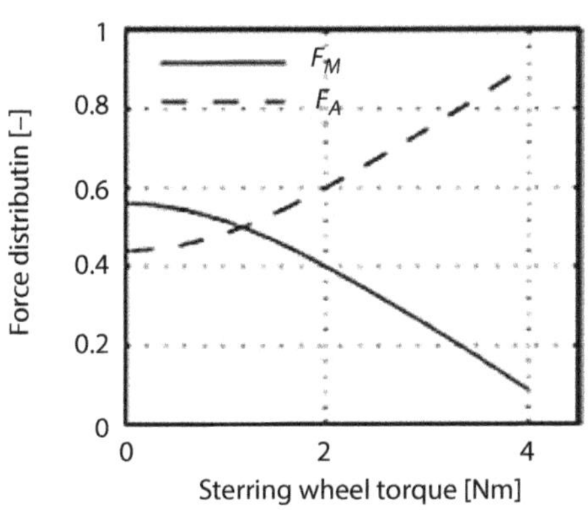

Figure 6.8 Force distribution.

$$F_A = \frac{F_{assist}}{F_{tot}}$$

Figure 6.9 shows the simulation of road disturbance with a sinusoidal input at two different working points where the disturbance is held constant at all conditions.

6.5.1.2 *Components of Hydraulic Power Steering*

The hydraulic pump is used to convert mechanical energy into hydraulic energy. The mechanical action of the pump takes the fluid from the reservoir in to the pump by producing a partial vacuum. This action also forces the fluid in the pump to go into the hydraulic system.

1. Types of pumps:

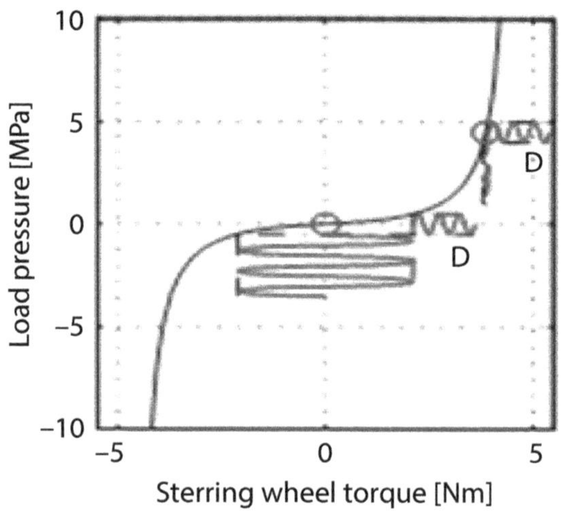

Figure 6.9 Disturbance propagation.

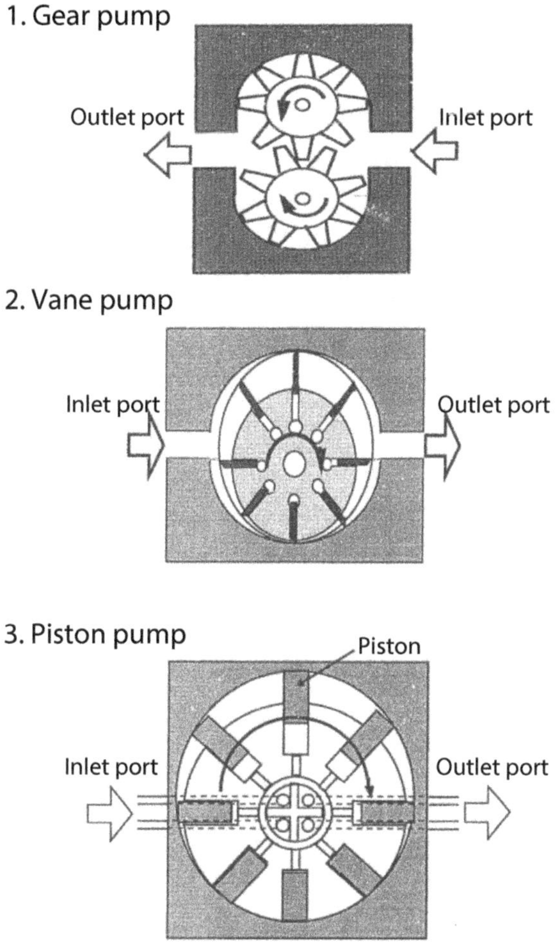

Figure 6.10 Pumps.

2. Hydraulic line:

Two types of line: high pressure line and low pressure or return line, e.g. Hose, Pipe.

3. Control valve:

A control valve consists of an inner and outer part. An inner part can be a spool, a piston, a ball, or a poppet, an outer part can be a stationary valve housing. The inner part is operated manually, electrically, pneumatically or hydraulically. The relative displacement of the inner part opens some valve ports while other ports are closed. There are different types of valves. Directional control valves are used to start, stop, and reverse cylinders and motors. Volume control valves are used to adjust fluid flow rate or speed of the load. The pressure control valves are used to adjust the force.

4. Power cylinder:

Cylinders are used to take the pressure or the power in a fluid system and convert it to mechanical force or power.

A cylinder consists of a cylinder body, and a rod attached to the piston. The cylinder body ends are sealed with oil seals. A cylinder of power steering system is generally integrated into the steering gear box.

5. Gear:

Rack and pinion gear mechanism and recirculating ball mechanism are commonly used for power steering systems.

6.5.1.3 General Design of Hydraulic Power Steering Systems

In the Figure below, a sketch of a HPAS system is depicted that shows pump, valve assembly, rack and

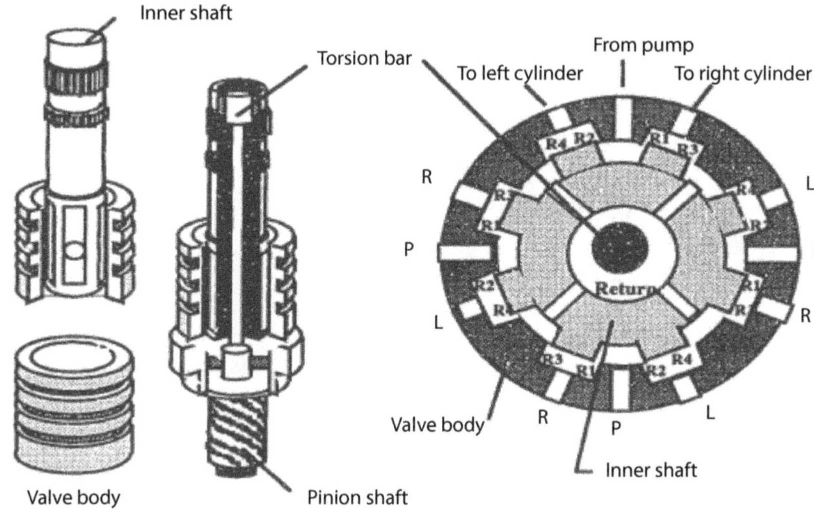

Figure 6.11 Rotary valve.

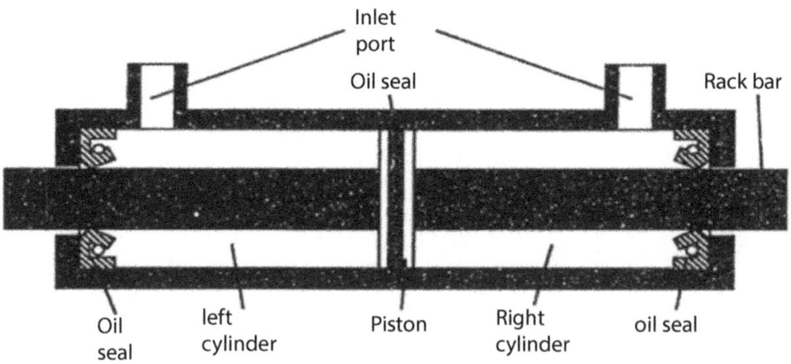

Figure 6.12 Cylinder arrangement.

the hydraulic cylinder. The most important part is the valve assembly having torsion bar in the core of the valve. The function of this bar is to activate the valve and at the same time transfer the applied manual force. The top part of the torsion bar is attached to the spool and the lower part is attached to the pinion.

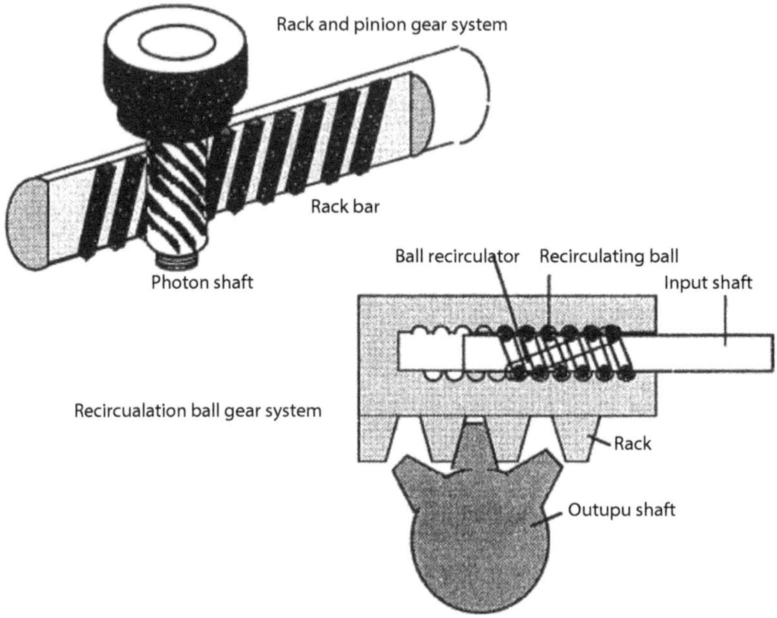

Figure 6.13 Gear.

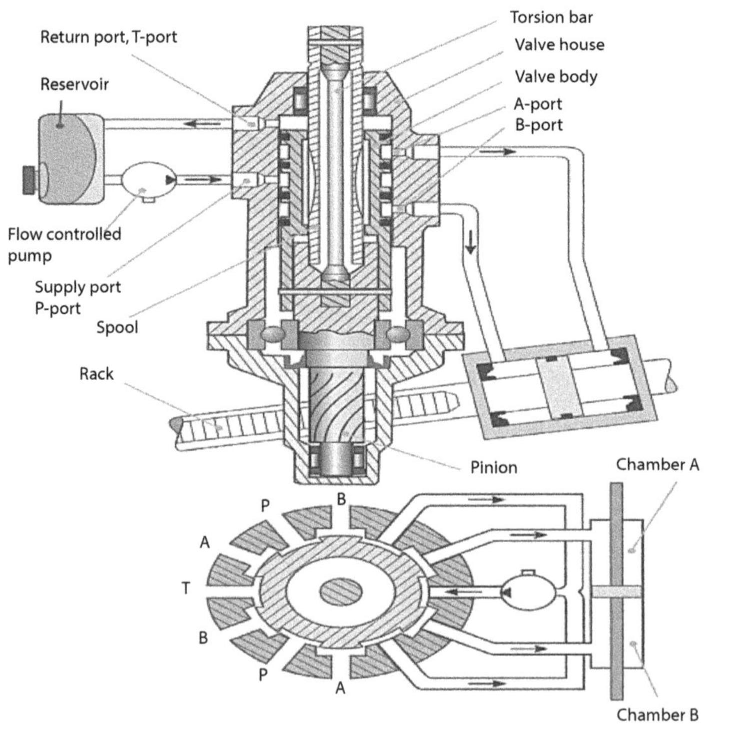

Figure of valve is shown below.

$$q_s = c_q A\left[-T_{SW}\right]\sqrt{\frac{P_S - P_L}{\rho}} + c_q A\left[T_{sw}\right]\sqrt{\frac{p_s + p_l}{\rho}}$$

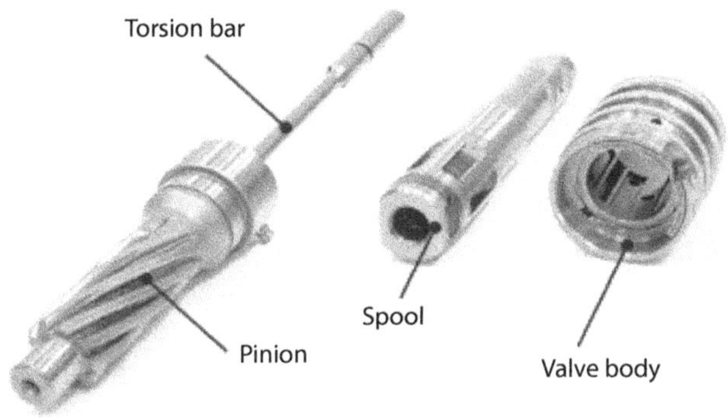

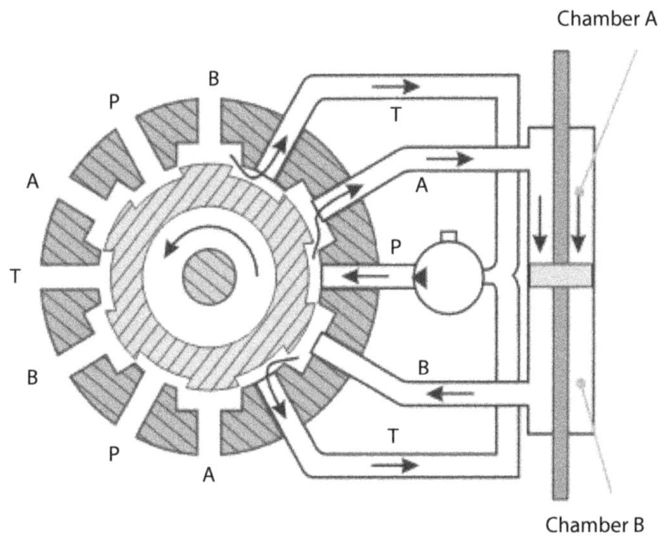

Figure 6.14 Valve displacement in a counterclockwise direction.

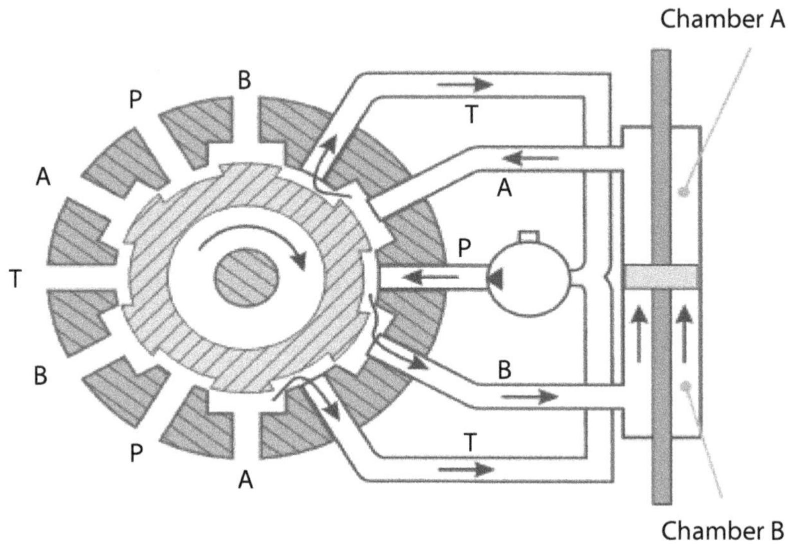

Figure 6.15 Valve displacement in a clockwise direction.

$$q_L = c_q A\left[-T_{SW}\right]\sqrt{\frac{P_S - P_L}{\rho}} - c_q A\left[T_{sw}\right]\sqrt{\frac{p_s + p_l}{\rho}}$$

Where,

q_s = pump flow	P_s = pump pressure
q_L = load flow	p_L = load pressure
T_{sw} = steering wheel torque	A = metering area
C_q = flow coefficient	ρ = density of fluid
	$x_r = l\theta_w$

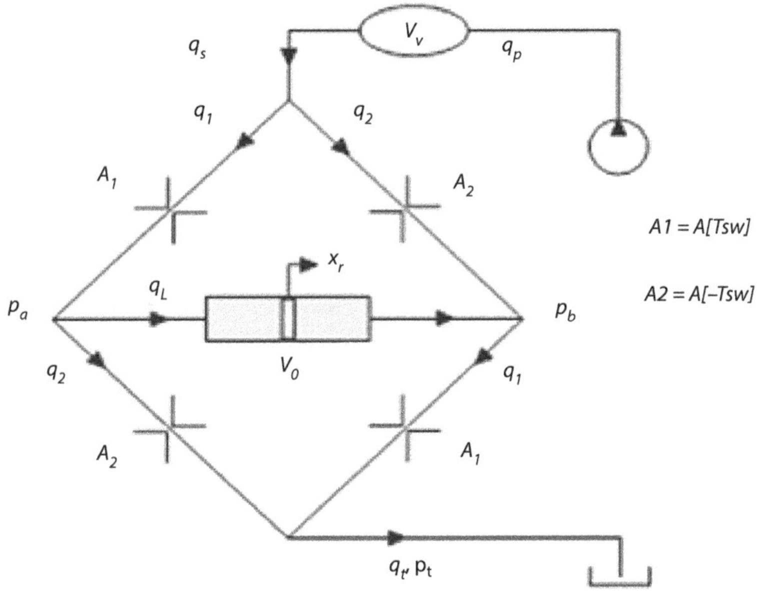

Where A_1 and A_2 are functions of the torsion bar angle.

Wheatstone Bridgerpresentation, where the multiple orifices are lumped together.

$$p_s - p_l = \frac{\rho}{4}\left(\frac{q_s}{c_q A[-T_{SW}]}\right)^2 \left(1 + \frac{q_L}{q_s}\right)^2$$

$$p_s + p_l = \frac{\rho}{4}\left(\frac{q_s}{c_q A[T_{SW}]}\right)^2 \left(1 - \frac{q_L}{q_s}\right)^2$$

The quotient of load flow and system flow is defined. This can be simplified to a normalized flow, q. In a normalized flow q = 1, defines the limit of the rack speed, with maintained ability to generate assisting pressure.

$$q = \frac{q_L}{q_s}$$

$$\dot{x}_{r_{\max}} = \frac{q_s}{A_p}$$

Solved Example:

<u>Assumption</u>

Weight of vehicle = 1600 Kg
Front axle weight = 950 kg
Total force needed to steer wheel = 600 N
Steering arm length (l)=0.1
Return pressure (N/m^2) =0
Pump flow rate (m^3/s) =0.0002
Area of piston (m^2) =0.001
Cylinder length (m) =0.15
Orifice flow coefficient = 0.6
Fluid density (Kg/m^3) = 825
Fluid volume (m^3) = $8.2 * 10^{-6}$
Fluid bulk modulus (N/m^2) = $5.5 * 10^8$

$$q_p = A_p \dot{x}_{rmax}$$

$$\dot{x}_{rmax} = \frac{q_p}{A_p}$$

$$\dot{x}_{rmax} = \frac{0.0002}{0.001} = 0.2\text{m/s}$$

$$\dot{x}_{rmax} = 0.2\text{m/s}$$

The steering wheel torque varies from $T_{sw} = (0\text{–}15)$ Nm

The normalized flow $q = \dfrac{q_L}{q_s}$ varies from (0–1)

Maximum pump pressure $p_{p_{max}} = (110\text{–}130 \text{ bar})$

$$p_L(T_{sw}, q) = \frac{\rho q_s^2}{8 c_q^2}\left(\left(\frac{1-q}{A[T_{sw}]}\right)^2 - \left(\frac{1+q}{A[-T_{sw}]}\right)^2\right)$$

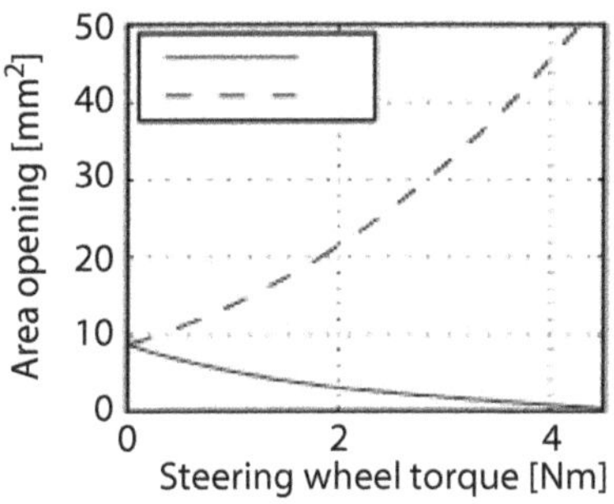

Area opening V/s wheel torque.

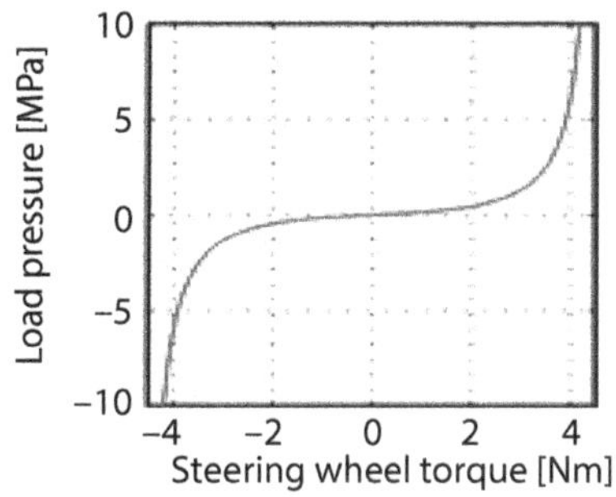

Generated load pressure V/s wheel torque.

To find the assisting force when the steering wheel torque is at maximum and when q=1

$$p_L(T_{sw}, q) = \frac{\rho q^2}{8 c_q^2}\left(\left(\frac{1-q}{A_2(T_{sw})}\right)^2 - \left(\frac{1+q}{A_1(T_{sw})}\right)^2\right)$$

$$p_L(T_{sw}, q) = \frac{825}{8 * 0.6^2}\left(\left(\frac{1-1}{20(15)}\right)^2 - \left(\frac{1+2}{30(15)}\right)^2\right)$$

$$p_L(T_{sw}, q)_{\max} = 11.31 bar$$

$$F_{tot} = F_{manual} + F_{assist}$$

$$F_{assist} = p_L A_p$$

$$F_{assist} = 11.31 * 10^5 \times 0.0004$$

$$F_{assist} = 452.4N$$

When steering wheel angle increases the steering wheel torque increases from 0 and also the load pressure also increases as the driver steers one of the metering area increases while the other decreases.

6.5.1.4 Hydraulic Power Steering Gear Design

Selection of material

Factorability and reliability were also considered. Based on this Aluminum T6 7075 were selected.

Module m = 3 mm (assume)
  Gear ratio G = 16 (assume)
  Pressure angle $\phi = 20°$
  Helical angle is from 20° to 45° (take $\alpha = 30°$)
  Number of pinion and gear teeth Z_p=7 teeth (assume)

Required

  Dimensions of pinion and gear
  Check the wear, bending, dynamic and other strength of gear

S. No.	Parameters or Dimensions	General formula	Pinion dimensions [mm]	Gear dimensions [mm]
1	Module (m)		3	3
2	Transverse pitch (Pt)	Pt = π*m	9.42	9.42
3	Number of teeth (Z)		112	112
4	Pitch circle diameter Do	Do = m*Z	21	
5	Addendum	0.8*m	2.4	2.4
6	Deddendum	1*m	3	3
7	Root circle diameter Dr	Do-2* deddendum	15	
8	Crown circle diameter Dk	Do + 2* addendum	27	
9	Depth of the tooth (h)	Addendum + dedendum	5.4	5.4
10	Depth of working tooth (h_w)	1.6*m	4.5	4.8
11	Tooth thickness (t)	1.5708*m	4.7124	4.7124
12	Fillet radius at root	0.4*m	1.2	1.2
13	Minimum clearance (c)	deddendum-addendum	0.6	0.6
14	Face width b	b = 1.15*Pt/tan α	18.7633	17.7633

Gear Parameter Analysis

$$G = \frac{Z_G}{Z_P}$$

$$Z_G = GZ_P = 16 * 7 = 112 \, \text{teeth}$$

- The normal and axial pitch of the pinion and gear can be determined from the geometric analysis of the figure below.

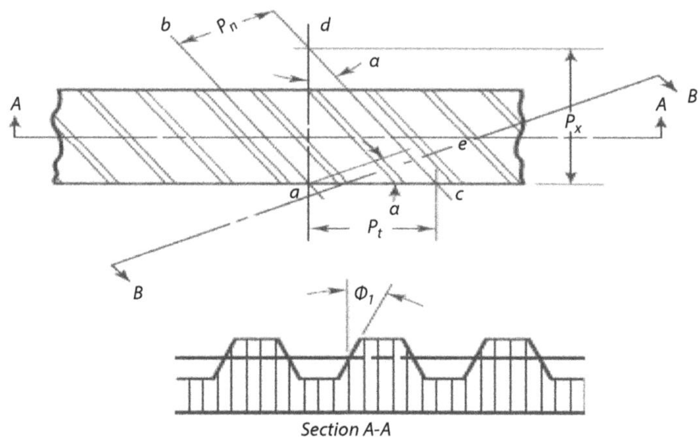

$p_n = p_t \cos\alpha = 9.42 \cos 30° = 8.158\text{mm}$

$p_x = pt/\tan\alpha = 9.42 \text{ mm}/\tan 30° = 16.316\text{mm}$

The transverse pressure angle $\phi_t = \tan^{-1}(\tan \phi n/\cos\alpha) = \tan^{-1}(\tan 20°/\cos 30°) = 22.8°$

(same for all gears and pinions)

Force analysis

Beam Strength of Gear Teeth – Lewis Equation

The maximum value of the bending stress (or the permissible working stress), at the section BC is given by

$I\sigma_w = M.Y$ $\qquad Y = t/2$

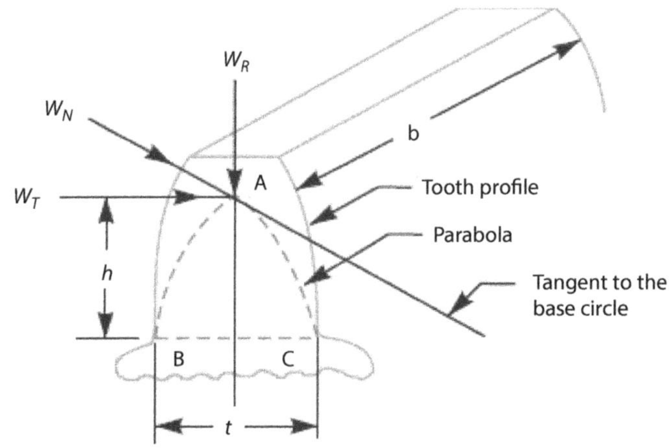

Where, M = Maximum bending moment at the critical section BC =WT×h,

I = Moment of inertia about the center line of the tooth = $b.t^3/12$,

By substituting these values into the above equation we get;

$$W_T = \sigma_w \times b \times \pi^* m^* y'$$

Lewis equation is applied only to the weaker of the two wheels. When both the pinion and the gear are made of the same material, then pinion is the weaker.

Permissible working (bending) Stress for Gear Teeth in the Lewis Equation

According to the Barth formula, the permissible working stress,

$\sigma_w = \sigma_o \times C_v$

Where σo= Allowable static stress, and

Cv= Velocity factor

$C_V = \frac{6}{6+v}$, for the peripheral velocities 5m/s to 10 m/s.

$$C_V = \frac{6}{6+5} = 0.545$$

σ_p =180.33

σ_w = 180.33× 0.545=98.36 MPa

600 = σ_w× 19 × π*3*0.175

σ_w= 19 Mpa The factor of safety F.s = 98.36 Mpa / 19Mpa = 5.3, therefore the calculated face width b is satisfactory.

Dynamic Tooth Load [3]

The approximate allowance for the analysis of dynamic loading effect can be calculated by using the velocity factor. The dynamic loads on the tooth can be due to the inaccuracies and irregularities of the tooth spacing and tooth profiles respectively. Along with that, the deflections of gear teeth under external loading can also cause dynamic load. A keen approximation to real-world conditions can be made by using equations based on extensive series of experimental tests, as follows:

$$P_D = P_T + P_I$$

Where, P_D represents total dynamic load, P_T represents steady load due to transmitted torque, and

Increment load due to dynamic action is denoted by P_I.

The increment load (P_I) relies upon the pitch line velocity of the gear, the face width, gear material, tangential load acting on teeth and accuracy of the cut made while gear generation. For average conditions, Buckingham equation is used to determine the dynamic tooth load on the helical gears, i.e.

$$P_D = P_T + P_I = P_T + \frac{21v * \left(bC\cos^2\alpha + W\ T\right)\cos\alpha}{21v + \sqrt{bC\cos^2\alpha + W\ T}}$$

Where, P_D = Total dynamic load (N),

P_T = Steady transmitted load (N), in determining the dynamic load (P_D),

The magnitude of tangential load (P_T) may be determined by neglecting the service factor (C_S) i.e.

$P_T = p/v = 1328.1N$, where pis in watts and v in m/s.

v = Pitch line velocity in m/s = 26.09m/s (from previous calculations)

b = Face width of gears in mm = 62.5mm (from previous calculations), and

C = Dynamic factor in N/mm = $C = \dfrac{K * e}{\dfrac{1}{Ep} + \dfrac{1}{EG}}$

Where K = Factor depending upon the form of the gear teeth.

= 0.111, for 20 ° full depth involute profile.

E = 200 GpaYoung's modulus (N/mm^2)

e = Tooth error action (mm) or module m (mm)

$$C = \frac{K * e}{\dfrac{1}{Ep} + \dfrac{1}{EG}}$$

$$= \frac{0.111 * 0.0125\text{mm}}{\dfrac{1}{200 * 1000\text{N/mm}^2} + \dfrac{1}{200 * 1000\text{N/mm}^2}}$$

$$= 138.75\text{N/mm}$$

Thus, the dynamic load P_D is

$$P_D = 1328.1\text{N}$$

$$+ \frac{21 * 26.09\text{m/s}\left(\begin{array}{c} 62.5\text{mm} * \dfrac{138.75\text{N} * \cos^2 30}{\text{mm}} \\ + 1328.1\text{N} \end{array}\right)\cos 30}{21 * 26.09\text{m/s} + \sqrt{\begin{array}{c} 62.5\text{mm} * \dfrac{138.75\text{N}}{\text{mm}} * \cos^2 30 \\ + 1328.1\text{N} \end{array}}}$$

$$= 6177.36N$$

The static tooth load on the tooth is given by

$P_S = \sigma_{es} * b * \pi\, m * y'$

Where σ_{es} represents the surface endurance strength (cast steel) = 616N/mm^2

b = 19mm face width

y' = 0.152, tooth form factor (from previous results) and

m = 3mm, (module of the gear)

$P_S = 616N/mm^2 * 19mm * \pi * 3mm * 0.152 = 5337N$

For heavy shock loading conditions $P_S > 1.5 * P_D$, So it is safe.

Finally, wear tooth load can be determined by using the relation,

ϕN = normal pressure angle = $\tan^{-1}(\tan\phi * \cos\alpha)$ = $\tan^{-1}(\tan 20° * \cos 30°) = 17.5°$

$$K = \frac{\left(616N/mm^2\right)^2 \sin 17.5}{1.4} \left(\frac{1}{200 * 1000N/mm^2} + \frac{1}{200 * 1000N/mm^2}\right)$$

$$= 0.815$$

$$P_{W(\max)} = \frac{DP.b.Q.K}{\cos^2 \alpha}$$

$$= \frac{210mm * 19mm * 1 * 0.815N/mm^2}{\cos^2 30}$$

$$= 4335.8N$$

Since the peak load for wear (4335.8N) is greater than the tangential load (600 N) on the tooth,, the design is satisfactory from consideration of wear.

6.5.2 Electric Power Steering

Electric Power Steering (EPS) is a full electric system, which reduces the amount of steering effort by directly applying the output from an electric motor to the steering system. It can solve the problems associated with hydraulic power steering [4].

6.5.2.1 *Working of EPS [5]*

The electronic control unit (ECU) of EPS analyzes the assisting power required based on the steering wheel torque applied by the driver along with the wheel position and speed of traction. The EPS motor reduces the steering torque required from the driver by rotating a steering gear.

As advancement in electronic steering, efforts are made to eliminate the mechanical linkage and convert to pure electronically controlled steering, which is called steer-by-wire. This works by transmitting digital signals to remote electric motors which in turn steer the vehicle.

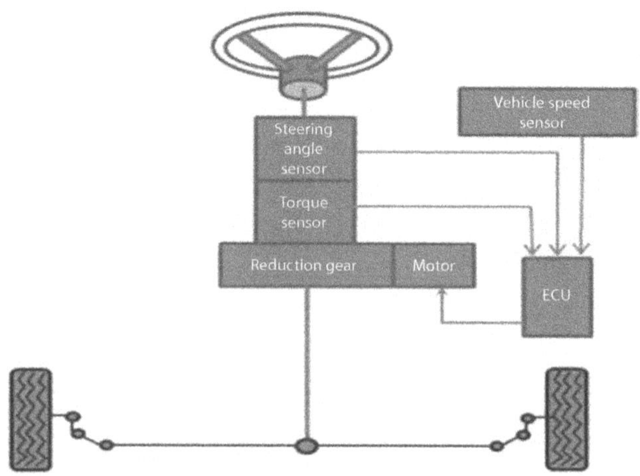

Figure 6.16 EPS arrangement.

6.5.2.2 Essential Components of an EPS System

EPS system designs have various components as:

1. Electric motor
2. Torque sensor
3. Angle sensor
4. Controller (CAN)
5. Vehicle speed and position sensor
6. Motor and steering mechanism coupling

1. Electric motors [6]

Electric motors are the most common electro-mechanical actuator in automobiles. Virtually every pump, fan and positioner in an automobile employs an electric motor. A typical automobile has about 25 electric motors and this number is growing with each new model. Luxury vehicles can have more than 100 motors.

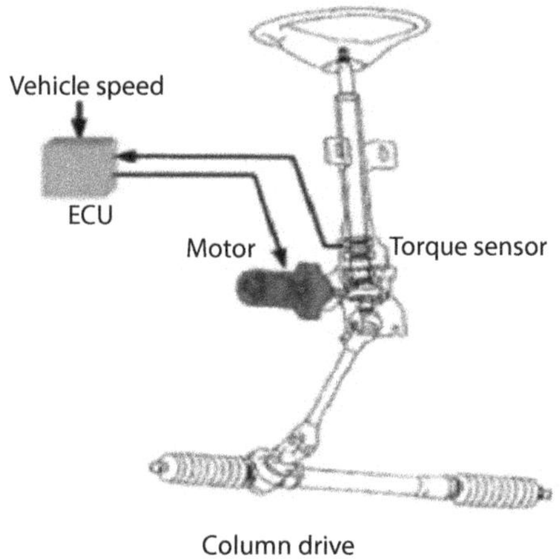

Figure 6.17 Components of EPS.

There are several types of motors including:

DC Motors - Brushed
DC Motors - Brushless
AC Motors - Asynchronous
AC Motors - Synchronous
Stepper Motors

2. Torque Sensor [7]

A torque sensor, in general is employed for monitoring the torque on a rotating system by converting torsional input into an electrical signal. Torque

Figure 6.18 Electric motor.

is measured by sensing the shaft deflection caused by a twisting force. Strain gauge is one of the most commonly used torque sensors to measure the torque applied to a shaft. Even though strain gauge provides high accuracy, their bulkiness and investment cost limit their application. Also the high levels of maintenance, makes strain gauge unsuitable for mass integration into automobile systems.

As an alternate, proximity and displacement sensors can be used to detect torque by measuring the angular displacement between the ends of the shaft. By fixing two similar toothed wheels at a distance apart on the shaft, the angular displacement can be measured [8].

3. Wheel Speed and Position Sensors (Vehicle)

Wheel Speed Sensors (Vehicle) measure the road-wheel speed and direction of rotation. These sensors provide input to a number of different automotive systems including the antilock brake system and electronic stability control. Wheel speed sensors

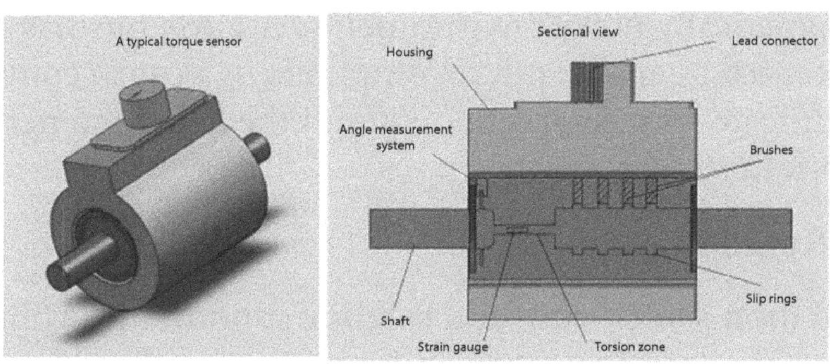

Figure 6.19 Torque sensor.

typically include a toothed (or optically encoded) shaft and a magnetic (or optical) sensor. The sensor counts the rate at which the teeth or marks pass by. Wheel speed sensors may monitor the crankshaft or driveshaft rotation in vehicles that do not need to know the rate at which individual wheels are turning. Otherwise, they monitor the rotation of the axle driving each wheel.

Position sensors

Position sensors are used in automobiles to determine the steering wheel position, pedal positions, seat positions, and the position of various valves, knobs and actuators. There are three main types of position sensors: angular, rotary, and linear.

CAN (Controller Area Network)[9]

CAN originally developed by Robert Bosch GmbH in 1986 as communication protocol between hardware and software for in-vehicle networks in cars. In CAN buses, twisted wire pairs were specially designed to withstand and work perfectly in electromagnetically noisy environments. A CAN bus has a potential to enable microcomputers in a car to communicate with each other without the help for a network host.

4. Rotation Angle Sensor

Rotation sensor converts rotation motion into voltage. The steering angle indicates where the driver wants to go. This information is also used by other systems such as the Electronic Stability Program (ESP), Active Front Steering (AFS), Adaptive Front Lighting (AFL), and so forth.

Coupling [10]

A coupling is used to connect two shafts for the purpose of transmitting power. The main purpose of couplings is to connect two pieces of rotating member while allowing some degree of misalignment or end movement. Shaft couplings are used in machinery to transfer power from one end to another end.

Calculation of Electrical Power Steering

EPS system has electric control unit (ECU) to calculate optimum assist torque based upon torque sensor and vehicle speed sensor, to output the calculation results to power unit. EPS system has electric control unit (ECU) to calculate optimum assist torque based upon torque sensor and vehicle speed sensor, to output the calculation results to power unit. Torque sensor is located between steering wheel and steering column and measures the applied torque by converting difference of twisted angle to electric signal. The motor is attached in steering column through reduction gear box.

The equation of hand wheel is like Eq. (1), and displacement, velocity and acceleration of hand wheel is zero by assuming fixed hand wheel.

$$J_{SW}\ddot{\theta}_{SW} + B_{SW}\dot{\theta}_{SW} + K_{SC}(\theta_{SW} - \theta_{SC}) = T_{SW}, \quad (1)$$

And, equilibrium equation of steering column is Eq. (2),

$$J_{eq}\ddot{\theta}_{SC} + B_1\dot{\theta}_{SC} + K_{SC}\left(\theta_{SC} - \theta_{SW}\right) = T_m - T_P, \quad (2)$$

where, J_{eq} is equivalent polar moment of inertia and B_1 is equivalent viscous damping of motor, reduction gear and steering column with respect to steering column.

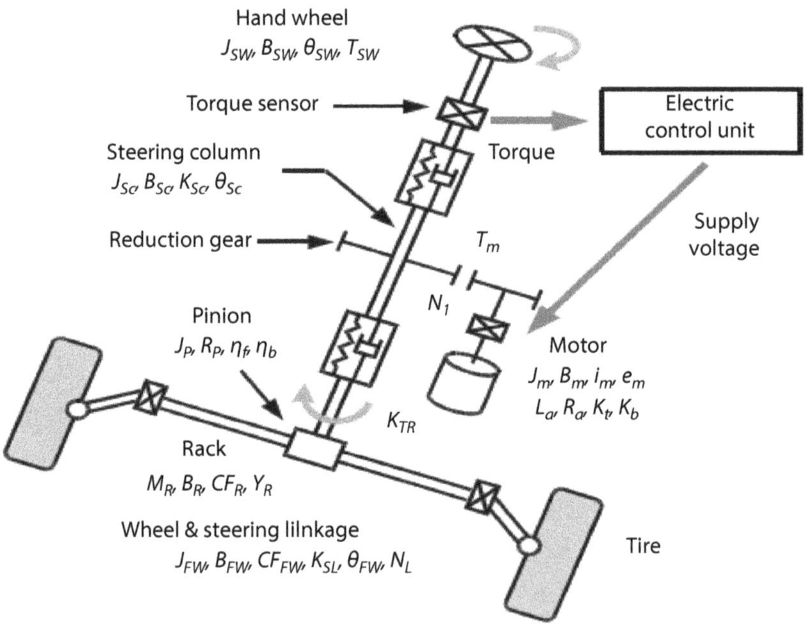

Where, J_{eq} is equivalent polar moment of inertia and B1is equivalent viscous damping of motor, reduction gear and steering column with respect to steering column.

$$T_m = \frac{N_1 K_i}{R_a}\left(e_m - K_b N_1 \dot{\theta}_{sc}\right), \quad (3)$$

The motor voltage in PD-control is Eq. (4)

$$e_m = -K_P\left(\theta_{SC} - \theta_{SW}\right) - K_d\left(\dot{\theta}_{SC} - \dot{\theta}_{SW}\right), \quad (4)$$

The torque at pinion in Eq. (2) can be assumed as Eq. (5).

By substituting Eqs. (3)-(5) into Eq.(2), Eq.(6) can be obtained as follows.

$$T_P = K_{TR}\left(\theta_{SC} - \frac{Y_R}{R_P}\right), \quad (5)$$

$$J_{eq}\ddot{\theta}_{SC} + B_{eq}\dot{\theta}_{SC} + K_{eq}\theta_{SC} = K_{TR}\frac{Y_R}{R_P}, \quad (6)$$

Where

$$J_{eq} = J_{SC} + N_1^2 J_m,$$

$$B_{eq} = B_{SC} + N_1 B_m + \frac{N_1^2 K_t K_b}{R_a} + \frac{N_1 K_t}{R_a} K_d,$$

$$K_{eq} = K_{SC} + K_{TR} + \frac{N_1 K_t}{R_a} K_P,$$

And, equilibrium equation of steering rack bar is Eq. (7),

$$M_R\ddot{Y}_R + B_R\dot{Y}_R + CF_R \times \mathrm{sgn}\left(\dot{Y}_R\right) = \eta_F \frac{T_P}{R_P} - 2\eta_B \frac{T_{KL}}{N_L}, \quad (7)$$

And, equilibrium equation of road wheel is Eq.(8),

$$J_{FW}\ddot{\theta}_{FW} + B_{FW}\dot{\theta}_{FW} + CF_{FW} \times \mathrm{sgn}\left(\dot{\theta}_{FW}\right) = T_{KL} + T_{ext}, \tag{8}$$

Here, torque at steering linkage is Eq.(9)

$$T_{KL} = K_{SL}\left(\frac{Y_R}{N_L} - \theta_{FW}\right). \tag{9}$$

By using Eqns. (1)- (9), driver's hand wheel torque T_{SW} is calculated from the input of road wheel load.

References

1. **Analysis of Davis steering Gear mechanism for Four Wheels and Six Wheels, Web address:** https://ijirst.org/Article.php?manuscript=IJIRSTV3I8030 Accessed on 18-04-2018.
2. **Hydraulic Power Steering System design in Road Vehicles, Web address:** http://liu.diva-portal.org/smash/get/diva2:23027/FULLTEXT01 Accessed on 18-04-2018.
3. **Beam Strength of Spur Gear Tooth, Web address:** https://srit-mech.weebly.com/uploads/2/1/9/8/21981046/beam_strenght_of_spur_gear_tooth-lewis_eqn.doc Accessed on 18-04-2018.
4. **Design and Full-car tests of Electric Power steering system, Web address:** https://link.springer.com/content/pdf/10.1007%2F978-0-387-77251-6_80.pdf Accessed on 18-04-2018.

5. **Electronic Power Steering, Web address:** https://cecas.clemson.edu/cvel/auto/systems/ep_steering.html Accessed on 18-04-2018.
6. **Electric motors in Automobiles, Web address:** https://cecas.clemson.edu/cvel/auto/actuators/motors-electric.html Accessed on 18-04-2018.
7. **Torque Sensors, Web address:** https://cecas.clemson.edu/cvel/auto/sensors/torque.html Accessed on 18-04-2018.
8. **Wheel Speed Sensors, Web address:** https://cecas.clemson.edu/cvel/auto/sensors/wheel-speed.html Accessed on 18-04-2018.
9. **Controller Area Network, Web address:** https://cecas.clemson.edu/cvel/auto/Buses/CAN.html Accessed on 18-04-2018.
10. **Coupling, Web address:** https://en.wikipedia.org/wiki/Coupling Accessed on 18-04-2018.

7

Hybrid Cars

7.1 Introduction

A hybrid vehicle (HEV) is a type of automobile which has a conventional power system with an energy storage system (RESS).

Mass production of these vehicles has been taking time due to charge time taken by regenerative braking.

7.2 History [1]

In 1901, Ferdinand Porsche designed a series-hybrid vehicle having a range of 50 km, speed of 50 km/h and a power of 5.22 kW.

Figure 7.1 Toyota's HEV.

Figure 7.2 The Escape Hybrid, launched in 2004.

7.3 Background

A more recent working prototype of the HEV was built by Victor Wouk and earned him the title as the "Godfather of the Hybrid". Wouk installed a prototype hybrid drivetrain into a 1972 Buick Skylark provided by GM. In 1989, Audi produced its first iteration of the Audi Duo (or Audi 100 Avant duo) experimental vehicle, a plug-in parallel hybrid based on the Audi 100 Avant Quattro. This car had a 12.6 BHPSiemenselectric motor which drove the rear wheels.

7.4 Production of Hybrid Electric Vehicles

Automotive hybrid technology became successful in the 1990s when the Honda Insight and Toyota Prius became available. These vehicles have a direct linkage from the ICE to the driven wheels, so the engine can provide acceleration power.

Figure 7.3 Hybrid vehicle sales chart.

7.5 Types of Vehicles

7.5.1 Motorcycles

Companies such as Zero Motorcycles and Vectrix have market-ready all-electric motorcycles available now, but the pairing of electrical components and an internal combustion engine (ICE) has made packaging cumbersome, especially for niche brands.

7.5.2 Automobiles and Light Trucks

Microhybrids is a type of small hybrid electric car. Diesel-electric hybrid vehicles may soon see mass-production.

7.5.3 Taxis

Figure 7.4 Hybrid-electric taxi.

7.5.4 Buses

Drive trains of buses consist of conventional diesel engines and gas turbines. A major issue for hybrid buses may still come from cheaper lightweight imports from the former Eastern bloc countries or China.

7.5.5 Trucks

In 2003, GM introduced a hybrid diesel-electric military (light) truck, equipped with a diesel electric and a fuel cell auxiliary power unit. Hybrid electric light trucks were introduced in 2004 by Mercedes Benz (Sprinter) and Micro-Vett SPA (Daily Bimodale).

7.5.6 Military Vehicles

The United States Army's manned ground vehicles of the Future Combat System all use a hybrid electric drive consisting of a diesel engine to generate electrical power for mobility and all other vehicle subsystems.

7.5.7 Locomotives

In May 2003, JR East started test runs with the so-called NE (new energy) train and validated the system's functionality (series hybrid with lithium ion battery) in cold regions.

Reference

1. **Hybrid electric vehicle,Web address:** http://taggedwiki.zubiaga.org/new_content/9d00b8c6d7038e9ea0197b2761ab1409Accessed on 18-04-2018.

8

Autonomous Cars

8.1 Introduction

Autonomous driving has recently garnered the attention of the media due to declaration of success by car makers, system partners and companies. Autonomous driving is synonymous with a concept from science fiction in many minds. Defining the term autonomous driving may help in establishing a more realistic picture of the developments that have to be made in achieving it. A peep into the present discussions on the use of autonomous driving reveals that a general terminology is yet to be defined.

Since the beginning of developments in autonomous driving technology, the word automobile was widespread among technologists. The term automobile was derived from a Greek and a Latin word and was first used when the car was invented. The word was formulated from the Greek word *autòs,* which indicates self, and the Latin word *mobilis,* which means mobile. Self-mobile is what is conveyed by the term automobile. It sure had created a sense of supreme joy as the driver did not need the aid of horses anymore. This term had a shortcoming in that it had failed to recognise the fact that the absence of horses had left the vehicle with a loss of autonomy. Horses were accustomed to obey the simple laws through dressage and instructions. Horses thus imparted a certain form of autonomy to the vehicles, which was lacking now.

This metamorphosis from horse carriages to automobiles had also spawned the loss of obstacle-avoidance skills and the occasional 'autonomous' rides were no more possible. In the instance of the driver being unfit to complete the journey, the horses would have safely brought the carriages to the usual destination. Even if that was not achieved, the horses would have left the vehicle in a 'safe state' by grazing along the wayside. An autonomous automobile aims to regain the lost autonomy and perhaps reach greater heights than in the entire history of the automobile.

8.2 Implementable Ethics for Autonomous Vehicles

Autonomous vehicles constantly interact with other road users which include pedestrians, bicyclists, and human or automated drivers. These interactions are governed by programmes and the order of importance given to the programmes by the programmers. The vehicle, just like any human driver, comes across a wide range of scenarios. From day to day common tasks like evaluating the distance from the vehicle ahead to unavoidable, unlikely events such as hatching the trajectory when an accident is inevitable, the vehicle must possess the ability to take the right decision in the right time. Standards of the society in which the vehicle runs arbitrate the behavior and control algorithms of the vehicles in such circumstances rather than statistics or test-based features.

The question whether automated vehicles without free self-will can be counted on for displaying moral behavior is yet to be answered. Still, the society will view the activities of the vehicles through the ethical lens. In the unlikely event of damage or injury being caused by the vehicle, the control algorithms of the vehicle will be under critical inspection and examination in a court of law. Even the daily tasks which involve social interactions undertaken by the vehicle will determine their degree of societal acceptance. Thus, the control algorithms framed by the programmers must follow legal and ethical rules and regulations.

Daily driving does not pose any major challenge as the autonomous vehicle can drive smoothly while ensuring all traffic laws are obeyed. On the other hand, dilemmas where it is impossible to meet all the constraints, though rare, cannot be excluded. A case where the car must cross a double yellow line to avoid collision with some other vehicle is an example where the vehicle cannot meet all the constraints yet must arrive at a best plan of action. In such cases a mathematical decision fulfilling all the constraints is infeasible. A solution to this would be creating a hierarchy of constraints with higher weight age to the constraints that cannot be violated compared to the other constraints. The vehicle will now be governed by the deontological constraints and when a dilemma arises, the best course of action will be adopted by following a consequentiality approach. The three basic laws of Robotics proposed by Isaac Asimov state:

1. A robot should not injure any one.
2. A robot must obey commands.
3. A robot must protect its self.

One of the key causes for deployment of automated vehicles is to decrease the number of fatalities and mishaps. Asimov's laws can be well adopted to frame three laws apt for autonomous vehicles due to their emphasis on protecting human life.

1. The vehicle must avoid collision with people on the road.
2. The vehicle must avoid collision with another vehicle.
3. The vehicle must avoid collision with any other object in the environment.

8.3 Mobility and Autonomous Driving

1. Valet Parking
   The vehicle's independence that enables it to move to the user from the parking lot and then drop the user at the set destination and park itself in a parking lot is the basic presupposition of the autonomous valet parking. The utilisation of autonomous valet parking would eliminate the effort that the driver is required to put in getting the vehicle from the parking space and leaving it back after a ride. This appears to be a door to door service to the users, similar to a taxi but different in the fact that the user would monitor the driving task. The gross time of travel would get reduced as the ingress and egress of the vehicle is made shorter. Various amenities as well as facilities could be improved, including number of seats, access to net, capacity, and multi-media. A larger fleet would provide the possibility of allotting different prices for different vehicles

equivalent to the existing establishment-based car-sharing and conventional car rental services.

Autonomous valet parking makes the need of stations unnecessary; however, terminals and stations could be set up to bring everything under one roof, which would increase the efficiency of vehicle maintenance. An advantage is that these terminals need to be situated close to customers but can be situated at a low-cost location. This could consume some time to reach the customer, which is regarded as 'dead' time since during the travel the vehicle has no other use. In an effort to reduce this dead time, various small terminals can be densely spread over the operating area.

2. Fully Automated car sharing with driver
   This mechanism assumes that the vehicle moves autonomously with drivers having the ability to drive independently. The benefits are lesser when compared to autonomous valet parking, from the viewpoint of car sharing. The only change this would bring in current car sharing is that it gives the driver the freedom of driving if he/she wishes to drive. On some routes where autonomous driving may be banned, for example, roads with many pedestrians, this case would

be advantageous. Such routes are mainly found in urban areas. On the other hand, in rural areas or areas far away from cities the uses of full automation of driving mechanism may be promising, However, this type of model offers less benefits compared to the previous use case.

3. Car-sharing methods
   This case supposes that the autonomous vehicle does not need any driver, even as a back up option. This eliminated the driver workstation, which opens up a whole new world of interior designing of vehicles. Now this vehicle will be viewed more like a compartment on a highway. This also opens up a platform that allows for a wide range of activities within the vehicle to be performed, like reading, chatting, playing, using of mobile phones and laptops, or even sleeping for a brief amount of time during the journey. Car sharing as vehicle on demand shows similarities to the existing taxi ride; however, it has the potential to replace taxis with its availability to a broader range of users.

8.4 Safety Concept for Autonomous Vehicles

The term safe state paints a picture of being secure, sheltered and guarded in one's mind. To someone else it might be be associated with low-risk. Thus, the term safe state is a relative concept and debatable, and necessary guidelines have to be framed. The ISO26262 standard said that an arrangement or operating mode of a system can be considered safe when there is an absence of unreasonable risk. When such a system falls below the brink of what the society defines to be risk free both in present and in the future, then the system can be deemed as a safe state. This brink of safety must propose a value at and beyond which the system would fail to fall into a safe state according to social, law-abiding and ethical aspects. The gravity of a personal injury and the chances of its occurrence together influence the risk factor.

Safe state of operation of a vehicles is defined as a state where risk is fair and reasonable. It is a very challenging task to define safe state since the allowable level of risk varies from person to person, which makes it tough to identify the threshold of safety. The scenario in which the automated vehicle is operated determines the level of allowable risks.

- All types of driving decisions
- Use of dynamic objects with autonomous vehicles
- Legal obligations
- The assignment and power capacity of autonomous vehicle

The scenarios described above on comparison with the threshold value will determine if the automated vehicle can be considered as safe or not. According to ISO26262, the chances of occurrence of injury and the gravity of that injury must be evaluated for every situation. Also, for every situation the solution must also be derived. This solution is yet to be found by researchers around the globe.

8.5 Conclusions

What seemed like a luxury in the last century has now become necessity. Similarly, though automated vehicles may appear to be a luxury and a not-needed technology to some, they have the potential to radically change the lives of people around the world. There are some major challenges to bring this technology into existence. One main challenge is framing laws and ethics. Once the ethics and laws are defined the entire technology would be upgraded. The next challenge is the safety aspects. Safety is a very ambiguous term. It must be clearly defined so that automated vehicles can be designed and programmed to fulfil the conditions specified to qualify as safe state.

Developments in the field of automated-type vehicles has produced a revolution in sharing by use of specific technology. Among these, autonomous valet parking proves to provide maximum benefits as far as car sharing is concerned. Autonomous vehicle technology is still in the developing stage as the question whether artificial autonomous vehicles have the capability to obey the ethics and moral obligations of society is yet to be answered. While there is no doubt that well-written programmes can make the autonomous vehicles follow laws and regulations, the range of situations that the vehicle may face is simply overwhelming. With more research and hard work, however, the solution to the current problems is not out of reach.

Index

Lightning Source UK Ltd.
Milton Keynes UK
UKHW021813270319
340000UK00003B/112/P

9 781119 479802